GERBERT GROHMANN

GERBERT GROHMANN

THE PLANT

Volume I

A Guide to Understanding its Nature

Translated by K. Castelliz
and
Barbara Saunders-Davies
from *Die Pflanze*, Vol. I

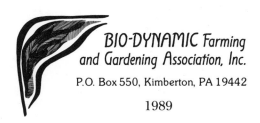

BIO-DYNAMIC Farming
and Gardening Association, Inc.

P.O. Box 550, Kimberton, PA 19442

1989

First published by Rudolf Steiner Press, London 1974

Translation of the fourth edition of
Die Pflanze, Erster Band, by G. Grohmann,
Stuttgart, 1959. Published by permission of
Verlag Freies Geistesleben, Stuttgart.

This edition published by
BIO-DYNAMIC FARMING & GARDENING ASSOC., INC.

Illustrations, execpt where otherwise stated, are
original drawings or photographs by the author.

ISBN 0 938250 23 X

Printed in U.S.A.

Contents

List of Illustrations

Gerbert Grohmann

Preface

The present book is the third edition of my first work which appeared in 1929. That little booklet was so eagerly received that a few years later, in 1933, I revised and enlarged it under the title of *Botanik*. During the war a new edition was required and I reverted to the original title of *Die Pflanze*. In the meantime the contents had so expanded that a single volume was not sufficient. The plant is a living thing and a book about it had to be living and capable of growth and transformation. So I took the section on non-flowering plants and combined it with most of the contents of my books *Metamorphosen im Pflanzenreich* and *Blütenmetamorphosen* into this volume.

The book was finished, printed and bound and ready for distribution when the unfortunate happenings in 1942 interrupted my writing activity. The whole edition was confiscated and went up in flames in Stuttgart without a single copy reaching the booksellers.

Fortunately, in spite of the obvious difficulties of printing such a work these days, it was possible to bring out a new impression, for many are waiting for the mental stimulus which can come from a study of the plant world. Besides scientists, I am thinking of such people as artists, teachers, doctors, farmers and gardeners. How many there are today who will welcome the opportunity to exercise their observation and thinking on the plant world! Our rigid hidebound times need such exercises as a medicine.

Years of study of Goethe's scientific writings and especially the researches of Rudolf Steiner, which cannot be valued too highly, were the guiding lines. But there would be no sense in writing yet another book on plants were the author not convinced that a new method of approach can convey new knowledge.

It is hoped the second volume dealing particularly with flowering plants will follow.

Finally I would like to thank all the friends who through conversations or their publications have helped me in my efforts.

October 1948   GERBERT GROHMANN

13

Foreword to the 1959 German edition

Dr. Gerbert Grohmann (5 June 1897–23 July 1957) died before he could prepare this new edition of his major work *Die Pflanze*. His plan to extend the contents – mainly by taking in the description of further plant families and additions to the cosmological viewpoint in Volume 2 – could not be carried out. So the present edition is simply another impression of the 1948 one.

Grohmann wanted to do more than just teach botany. He found a new way of looking at plants, a way which leads to a living comprehension of their varied and ever changing forms and to an understanding of the laws which shape them. This book is an important contribution to the more living exploration of nature now possible on the basis of Rudolf Steiner's Anthroposophy.

The many editions, often augmented, which have appeared since 1929, bear eloquent witness both to the approval met with by Grohmann's botanical method, which takes into consideration the spiritual aspects of the plant, and to its wide impact. We hope it will continue to find such approval in the future.

Dr. F. A. Kipp

_I_ _Looking at the Flowering Plant_

_I_ _Leaf, Root and Flower_

The most characteristic part of a plant is the green leaf. In its thousand-fold variations we recognize the leaf as the primal organ which is the living link between earthly and cosmic life. Goethe's words: 'How would I know that this or that form was a plant were they not all designed from a single pattern?' apply particularly to the green leaf. It is attached to the stem and the point from which it springs is called a node. Leaf, node and stem form the basic unit from which all plants develop.

Fig. 1.

In the axil of the leaf is the bud, and if it starts to grow it becomes a new plant, a branch, capable of carrying its own flower. The stem is a linear structure which raises the plant up towards the sun. The leaf is a plane, a surface, and with it the plant can catch the maximum light. The bud is a contraction of the plant into a point where all future possibilities lie dormant. The capacity of regeneration – the possibility of expanding and contracting and expanding once again at rhythmic intervals – is characteristic of the purely plant nature. To follow up the development of stem, node, leaf and bud means to listen in to the secrets of vegetative life. One learns how nature sets about the creation of her children, however varied be their shape. The repetition of events in fixed rhythms has its limits set at both ends of the plant. Every plant shows us that its leaves will unfold only under the life-giving light. Below the earth's surface are no green leaves; the subterranean shoots are pale and undefined in shape, while above, in the flower, the processes are so different that some of the organs cannot easily be recognized as metamorphosed leaves.

Let us first look at the root. It can take many different forms. Sometimes it is finely branched, sometimes it consists of many single strands as in bulbous plants. It can also be swollen like a turnip or drawn into a tap root like a dandelion. The root grows in the opposite direction from the shoot, thus showing its affinity to the earth. It is averse to light and seeks darkness and damp; only rarely does it appear above ground. Through its root, the plant becomes united with Mother Earth. Though each species has its characteristic root system, the differences between plants are far greater in their upper parts. Living in the soil gives roots a certain similarity of appearance.

The root has two main characteristics. First, it has great vitality and regenerative capacity, and spreads easily. When a seed germinates, it develops a radicle or rootlet before a shoot. Without connection with the soil, life in the light would not be possible. Only the finest and outermost root tips (hair roots) can perform the vital function of taking up water and soluble mineral salts. When we transplant a plant it will wilt until it has been able to renew its hair roots. The totality of the delicate root tips comprises the actual living root system. We must look at the root in a different way from that part of the plant which is above ground. The stem grows towards the sunlight vertically; the root spreads out spherically in the soil. Although it consists of single strands, it is a spherical structure. It has the same tendency as a drop of water. By contrast the stem can be seen as a living ray of sunlight reflected by the earth. The plant is terrestrial as regards its root, cosmic as regards its stem. Between these two poles stands the two-dimensional flat leaf which in its shape portrays a conflict between spherical and linear forms.

The second point about the root is that it tends to harden very quickly. Although it is the most vital part of the plant, the vital areas are shortlived. It remains young only by the continual growth at its tips. Thus enormous woody perennial rootstocks can develop. Through its hardening tendencies it becomes related to the mineral.

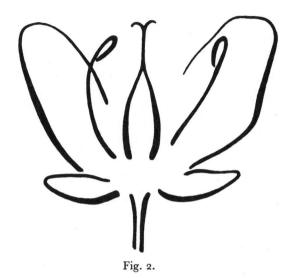

Fig. 2.

The metamorphosis of the flower is different. Here formative forces take hold of the leaf and transform it completely. It loses its green colour, no longer serves the purpose of light-absorption (photosynthesis) and sometimes takes on strange shapes. The leaves are arranged along the stem at rhythmic intervals as it grows. In the flower the leaves stand tightly packed in whorls. Sepals, petals, stamens and pistil originate from the centre in a star-like formation. This concentration of parts has a very significant result, for the flower becomes a unified organ. Each part is subservient to a superimposed whole. The flower can be considered as a plant of a higher grade whose leaves, gathered around a common node, rise above the purely vegetative sphere. Colour and scent appear, while the vitality of the rest of the plant is reduced. There are no buds in the axils of petals or sepals. A closed flower bud hides the interior of the blossom like a secret. It is a gesture reminiscent of animal life, where the organs are also hidden and enclosed within the body. The shapes of some flowers are also reminiscent of the animal kingdom. Many resemble lower animals, especially insects, or parts of them such as faces, snouts, mouths with lips, teeth, beards etcetera. It is a peculiarity of vegetative life to copy the gestures of ensouled creatures, but only outwardly – there is no inner correspondence. In the process of flower development, however, the plant does touch a soul-life element.

Colour itself is indicative of this. Rudolf Steiner says: 'Colour is the soul of Nature and the Cosmos, and we become aware of this soul when we experience colour.' Of all parts of the plant it is the flower that speaks most intimately to us. Something related to us seems to meet us in their colours and shapes, sympathy and antipathy change with our moods and children have different preferences from adults. Flowers are also often used as symbols.

This fleeting inclination towards the animal kingdom is reversed in blossoming. Parts that were concealed are spread out and exposed to the light. The corolla is an open secret – still an organ, but open to the traffic of bees and butterflies, and to the wind. The plant when in flower is raised a step on Nature's ladder, but, though touching the animal world, it has nevertheless opposite tendencies. It exposes itself to the outer realm of light whereas the animal turns itself inwards.

2 Germination and Propagation

The seedling is the first stage in the development of the flowering plant from a seed, but it is a definite phase which is not part of the real development of the plant itself. The plant pushes the seedling out in front of it as a kind of advance guard when it wants to establish itself on the earth. The moment the radicle has broken through the skin of the seed and turned downwards, the cotyledons come up to the light. They become green very quickly and can look almost like leaves. However, by their nature they are something totally different from leaves. They have simple shapes – round, egg-shaped, heart-shaped or narrow linear, but without mid-rib or serrations – and differ clearly

Plate 1: Seedling of *Chelidonium major* (Greater Celandine); cotyledons and first true leaves. (Enlarged).

from the leaves of the stem. The first true leaf looks very different. However small and contracted it may be, it already bears the mark of its species. To tell a plant by its cotyledons requires very specialized knowledge. According to Goethe's terminology, the seedling is the most primitive form of the archetypal plant as it has only one node. Every plant conforms to this kind of universal pattern when it germinates from the seed. Even the seedling's radicle is only a preliminary stage; it dies off and is replaced by lateral roots of a different structure. If a tap root is formed the original model is transformed.

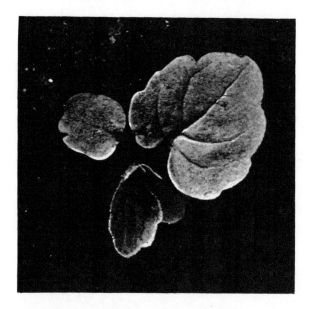

Plate 2: Seedling of *Valeriana officinalis* (Valerian); cotyledons and two true leaves (Enlarged).

Since the cotyledons are not true leaves they do not fit in with the pattern of leaf metamorphosis. The plant takes a leap from the cotyledon stage to its first leaf. There is no continuity. If a plant's leaves are, for instance, serrated, the cotyledons will still be simple and primitive, and if a plant does not develop any leaves, as in cacti, a normal seedling appears first, then a minute cactus grows up between the cotyledons and gradually becomes a plant of considerable dimensions.

We must note that the seedling does not develop in the soil. It is already contained within the seed. We can find it there tightly

22

compressed, sometimes rolled or folded up, but existent in all its essential parts. Some plants do not push it up above the soil but keep it within the seed (oak, horse-chestnut, pea). Then the cotyledons are fat and fleshy, quite unlike a leaf, and serve as stores for starch and protein. Without doubt, the seedling is a repetition of a previous stage of development during which the plant was an altogether different living organism.

The embryo seedling is created in the ovary of the parent plant. In this ovary it is protected from all terrestrial influences, even from the

Plate 3: Seedling of Beech; cotyledons and first true leaves.

outer sunlight, by the carpel leaves. It is as if it were in a cosmic enclave where the new life was conceived. This enclosed ovary is the signature only of the higher flowering plants. In the conifers it is not yet there.

To proceed further we must now consider the laws governing the phyllotaxis of the leaves. Every casual nature-lover knows in how many different ways the leaves can be arranged. The Stinging Nettle shows a simple arrangement. The leaves stand in pairs. In the majority of plants, however, they are arranged in a spiral which winds its way up the stem. In the phyllotaxis of leaves there are hidden laws characteristic of the individual plant species. In the rose, for instance, we have to move two-fifths of the way around the stem before encountering the next leaf. It is the sixth leaf that stands exactly above the first one.

A lot of work has gone into the study of these laws, but they have remained an enigma and will continue to do so as long as the cause is looked for in the plant itself. Rudolf Steiner pointed out that phyllotaxis is a mirror image of those cosmic laws which regulate the movement of the planets. Hence we have to look to the cosmos if we want to understand the form of plants. Among other things it has been found that, in many cases, the position of the young leaves at the growing point is different from that which they eventually occupy on the stem. Thus the cotyledons of the dicotyledons stand in pairs or crosswise in pairs, but mostly their phyllotaxis is much more complicated. Such arrangements are achieved by the plant through different rates of stretching of the parts of the stem between the leaves.

What does it mean when a plant starts off with a different leaf-arrangement from the one it finishes with? As it grows, it pushes its seedling form ahead of it. Every node is a repetition of the seedling stage, and only in second place do the specific formative forces of the species assert themselves. There are thus two different streams working in the plant: one that makes the material malleable and one that moulds it. The seedling is the pliable foundation on which the formative forces can work. These formative sculpting forces are closely related to light. Light works as a formative force. The astronomer Kepler makes the interesting reflection: 'It is the strongest tie which binds the lower realms of Nature to the heavens . . . in these lower earthly realms there dwells something spiritual capable of geometrizing, which is quickened through the geometric, harmonic relationship with the celestial rays of light. Whether – like the earth globe – all herbs and animals have the same capacity, I cannot say. It is not unthinkable.'

Having looked at germination from seed, let us now turn to vegetative propagation. In the preface to his book *The Metamorphosis of the Plant*, Goethe writes: 'Every living thing is a plurality, not an individual. Even if it looks like an individual it is still an agglomeration of living, independent beings – beings identical in their potentialities even though in appearance they may be similar or dissimilar to a greater or lesser degree.' He states further on in the same book: 'There is no doubt that the plant, even a tree, consists of individual units which resemble each other as well as resembling the whole, which appears to be an undivided individual.' We do not look upon a branch as a self-contained plant, but neither is it a new creation when we cut it off and make it grow roots as a cutting. Each node has the inherent ability to grow roots, as we can easily see when observing plants with recumbent stems. Propagation is simple then. The stem between the nodes loses its function, and the tips of the shoots become separate little plants growing out of the nodes. The word 'reproduction' cannot be used in the sense of multiplication in such cases. One is

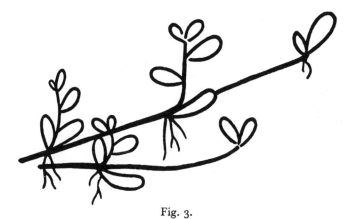

Fig. 3.

tempted to call it just growth, for where and how a plant attaches itself to the ground is often of only secondary importance to the living whole. The whole plant species, not a single plant, is the unity; it is responsible for the life of the individual plant. Vegetative propagation is used widely by the gardener, as for example in the case of the potato. The seeds of the green, tomato-shaped fruits are used only for breeding new varieties. Normally the seed potato itself is put into the soil. The potato is a stem, a thickened, subterranean, creeping stem filled with starch. If it were a tuber it would have no buds or eyes. Being part of the stem, it becomes green when exposed to light. It develops shoots even before it is put into the soil. We can cut a seed potato into pieces and each piece will become a potato plant, provided it possesses an eye or bud.

25

Among the highly-bred tropical plants are some which no longer develop any seeds (some bananas for instance). Such plants can be propagated only by cuttings. The sugar-cane has almost entirely lost its capacity to bear flowers. Cuttings of it, however, strike easily. Again consider all the pot plants of foreign origin which will flower but do not develop seeds. They are always propagated vegetatively. If we divide a fuchsia plant into as many pieces as we can, grow all of them till they flower, they still belong together; they are one, like the branches of a tree. However often you strike a new cutting is of no importance. This is the secret of vegetative life: it is divisible and yet remains whole. The plants can be a whole even when they are divided and the parts have become independent.

3   Leaf Metamorphosis

After these general observations on the nature of plant life, we will now consider individual cases which illustrate the variations and gradations in which the 'archetypal' appears in its earthly manifestations.

The leaf can be formed and transformed in every conceivable way. A single form alone is meaningless; it can be understood only in relation to other forms in the flow of metamorphosis.

The plant is a mirror of oppositions. The polarities to be found in the upper and lower parts of the plant are incomprehensible unless viewed within a much wider context. The way in which these opposites interact. the enhancements, transitions or even leaps that occur, are what make the individual characteristics of a plant species. Above all the plant is a picture of its external conditions – wetness, dryness, light, shade, soil, etcetera. Spring vegetation expresses different laws from that of high summer.

As starting point let us choose a type of plant which expresses itself particularly strongly in the metamorphosis of its leaves: the *Ranunculacea*. They have the advantage of being widely known and grow almost everywhere, so the reader can pursue his studies in the field. In fact we would recommend it, as only in this way can one get real pleasure and satisfaction from the study of plants.

If we compare various species of the buttercup family, we cannot fail to notice certain common characteristics. The 'type' is clearly evident, yet each species has its particular distinctive qualities.

Ranunculus acris (Field Buttercup)

In spring, as soon as vegetative life re-awakens, the Field Buttercup shoots up from its old root stock. In our study of the metamorphosis of the leaf, we will take the leaf development of only one season of growth, so we will ignore the leaves following the seedling stage for the present.

All the leaves shown in Plate 4 are from a single plant and only from the main stem; the side branches are ignored. In this way we can see a continuous development, the stages of which can be directly compared. The same principle has been followed in the subsequent series of leaves.

Looking at the series of leaves as a whole, we find first an expansion, then a contraction of the single leaves as we follow them from the root to the top of the stem. The leaves of the middle region are not only the largest, they also have the longest stalks, and their tips and blades are the most elaborately formed. The first four leaves form a rosette at the base of the stem. With the fifth, we have a leaf springing from the stem and it shows clearly the transitional stage.

Another contrast between below and above appears in the forms themselves. We see how the 'crowfoot' shape of the leaves develops out of a less typical beginning. The first leaf is the most entire one, giving the impression of a round leaf, while the following ones are more polygonal.

As they near the upper pole of the plant the forms again become simpler. The uppermost leaves are like single sections of the middle ones. At the same time they become narrower and look as if the leaf stalk had been pushed into the blade and the two had become one single organ.

The plant does not return to its original form. It disappears more and more from the world of space. Contraction and tapering are outer signs of a transformation. The plant is being extinguished in order to emerge on a higher level. The blossom is anticipated.

Ranunculus bulbosus (Bulbous Crowfoot)

This plant owes its name to the bulb-like formation at the base of the stem. It is also recognizable by the calix which is folded backwards against the stem. The leaf metamorphosis of the Bulbous Crowfoot differs from that of the Field Buttercup in that the continuity is interrupted. Compare the two top leaves with the two bottom ones and then look at the middle leaf. It gives the impression of being assembled from two diverse halves. The extremity pushed out by an extra piece of stalk repeats the shape of the lower leaves, while the pointed leaflets issuing from the leaf base already presage the future development. The 'above' and the 'below' meet but do not interpenetrate.

Plate 4: *Ranunculus acris* (Field Buttercup); sequence of leaves.

The shapes of the upper and lower leaves originate from opposite form principles. Looking at the lower leaves, one can see that the indentations and concavities appear to be eaten into from the outside, from the periphery – a centripetal action. With the upper leaves it is the reverse. The forms show the radiating tendency of the mid-rib and leaf veins, thus expressing a centrifugal force. It goes without saying that in none of the leaves is one or the other principle working *exclusively*. In the upper plant, however, radiating tendencies predominate; in the lower plant, constricting ones.

Ranunculus auricomus (Wood Crowfoot, Goldilocks)

Here the leap in the leaf metamorphosis is still more pronounced. The plant seems to be made up of two halves which have nothing whatever to do with one another. The uppermost leaves clearly show the radiating tendency of the flower.

The study of leaf metamorphosis leads to an understanding of the quintessence of the plant kingdom – dynamics. We shall never be able to grasp life if we stare at the fixed shape, at the finished product; rather we must consider the process of 'becoming'. The single leaf is only a mile-stone in the plant's development, a visible product left by life flowing on. Anyone who would try to explain a later leaf from an earlier would make the mistake of taking for a cause what is already a result.

The metamorphosis which eventually leads to the development of the flower is already announced by the transition from the first leaf to the second one. The flower is the aim of all the changes; it is the cause of all the profound transformations to which the herbaceous plant is subjected. If the vegetative process alone existed, the plant would grow on forever unchanged, as can be observed in exceptional cases. One instance among the flowering plants is *Veronica beccabunga* (Brooklime), a plant that grows in ditches and at the edge of ponds. Wherever the stem lies flat on the ground, its nodes send forth new roots whilst the terminal shoot grows on and on. Brooklime is a plant of luxuriant growth. Out of the axils of the pairs of leaves grow the panicles of flowers. Compared with the vitality of the shoot, the flowers are but poorly developed, and their impulse is not strong enough to check the

Plate 5: *Ranunculus bulbosus* (Bulbous Crowfoot); sequence of leaves.

Plate 6: *Ranunculus
auricomus* (Wood Crow-
foot); sequence of
leaves.

32

Plate 7: *Veronica beccabunga* (Brooklime); typical example of a plant which does not stop growing when the flowers appear. Hence there is no metamorphosis of the leaves.

33

growth. The vegetative life flows on unimpeded, overpowering the flowers, as it were, and passing them by.

So we can see that the floral impulse is needed to bring about the metamorphosis of the leaf in the higher plants. If it is missing or weak, vegetative life goes on unchanged. It is the same in plants which Nature has raised only as far as the leaf stage, for instance, the Ferns. The fern gets as far as the leaf and, however wonderfully it may shape its green organs, subdividing and re-assembling their parts, there is no question of metamorphosis as we know it in flowering plants.

The transition from leaf to flower is not a continuous one. Before calix and corolla can appear a point must be reached at which the plant shrinks almost out of spatial existence. It is a law of Nature that all development is discontinued, and it was an ill-starred idea of science that Nature does not make leaps, an error that has brought many misunderstandings in its train. Wherever Nature introduces anything new she takes a leap. Higher stages spring from apparently nothing. As the plant prepares to flower, room has to be made for the new organ. The idea of the flower cannot be inferred from looking at the green parts of the plant; it appears as a surprise, as a miracle, endowed with quite new characteristics. No one could foretell it unless he had seen it before. Until the flower emerges, it is wrapped in the veil of secrecy. Plates 8–13 illustrate a few more instances of leaf metamorphosis.

Plate 8: *Ranunculus arvensis* (Corn Buttercup), an annual weed on cultivated land. The sequence of leaves from cotyledons to flower is continuous, without sudden jumps.

Plate 9: *Delphinium consolida* (Field Larkspur), not a Ranunculus yet belonging to the same genera. Another annual.

Plate 10: *Scabiosa columbaria* (Small Scabious). The Scabious family has opposite leaves. What was of interest in the Bulbous Crowfoot becomes particularly obvious here: the divisions start on the lower part of the leaf and move gradually upward. The uppermost leaves are quite unlike the first ones.

Plate 11: *Knautia arvensis* (Field Scabious). This plant shows the same law as the previous example. The leaf begins to divide at the lower part nearest the stalk. The leaves on the right side show a curious constriction which continues through several stages.

38

Plate 12: *Lactuca muralis* (Wall Lettuce). Even a superficial view of this wonderful woodland plant shows that here the leaf metamorphosis is of a more hidden nature. The bottom and topmost leaves have nothing in common. Is it possible to find Goethe's law of polarity and enhancement even in this plant? Beginning with the fifth leaf, the stalk undergoes a remarkable transformation. It widens more and more, whilst the original leaf itself wanes and in the end disappears altogether. The two uppermost leaves are just clasping stalks encircling the stem, and pointed at the end. So there is a relation between the top and bottom leaves even though they are so dissimilar.

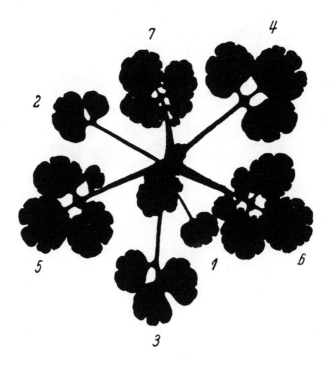

Plate 13: Young plant of *Chelidonium major* (Greater
Celandine) with its first leaves, from the first simple
undivided leaf to the lobed ones (7) through all the
intermediate stages.
(Natural size).　　　　　　　(From a herbarium specimen.)

4 *The Prototype of the Flower*

The parts of which the flower is composed are similar in all plants: calix, corolla, stamen and pistil. However varied their forms may be in individual cases, their identity is unmistakable, nor is their succession accidental. The calix always forms the outermost circle, then follow the petals with the stamens; the pistil – whether monogynous or polygynous – is the central part of the flower. There is no exception to this rule.

Here we are confronted with a law in plant organization which should not just be taken for granted. Could it not be that behind this rigorous norm significant secrets are hidden?

The perianth is composed of calix and corolla. In the Crowfoot (Buttercup) both circles are five-membered. If a line is drawn connecting the tips, two regular pentagrams will result. If the flower is turned upside down, one can see that the petals and sepals are placed alternately. The yellow petals grow in the gaps between the pointed sepals.

During the bud stage, the green sepals are tightly closed. They conceal the inner parts of the flower. Only at the moment of flowering do petals, stamens and pistil become visible. The petals of the buttercup can be pulled off individually, as they grow separately right down to the base. The buttercup is therefore polypetalous. It differs from those flowers where the petals are united at the base, such as Bluebell, Convolvulus, Cowslip, etcetera. When it withers, the petals, stamens and part of the calix drop off. Only the peduncle with the pistil remains, for the seeds do not ripen till a later stage of development.

Goethe, in *The Metamorphosis of the Plant*, says of the development of the stamens: 'And so an androecium arises, when the organs which were hitherto seen expanded as petals re-appear in a highly contracted and, at the same time, a highly refined condition.' The leaf, which has to be transformed, is rolled in from the edge and as end-product we get the stamen, i.e., anther and filament.

Plates 14 and 15: Flower of *Ranunculus acris*, (Field Buttercup). (Enlarged).

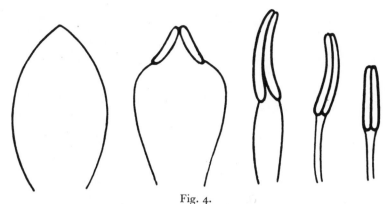

Fig. 4.

Transition from petal to stamen. (White Waterlily).

In order to see this metamorphosis more clearly, let us take the white water-lily, where the transformation of the petals into stamens occurs in stages, so that several different phases can be observed simultaneously. 'A true petal, little changed, contracts in its upper margin, and an anther, in connection with which the rest of the petal takes the place of a filament, makes its appearance' (Goethe, *The*

Plate 16: Red Waterlily. The stamens are erect and resemble a second corolla.

Plate 17: Double Marsh Marigold. As the stamens are transformed into small petals, their spiral arrangement is clearly seen.

Metamorphosis of the Plant). Such intermediate stages appear in many flowers. These shapes can be produced artificially in double flowers. This shows that there is an inner connection between petals and stamens, for they have many properties in common, such as delicacy, colour, etcetera. The filaments are often attached to the inside of the petals and therefore drop off with them at the same time.

One can see how precise is Goethe's statement about stamens being greatly contracted and yet highly refined organs, when one appreciates the contradiction this implies. How is it possible for an organ to contract without at the same time getting thicker and harder? Is not pollen the most delicate product of the plant? This contraction is not produced by external factors as, for example, in alpine plants. It is the expression of an entirely inner metamorphosis. At this point the plant only *seems* to be extinguished. Structurally the plant shrinks. but *functinonally* it expands immensely when its pollen is dispersed. The plant's formative force can be perceived in more than what is contained within its 'skin'. Whether the pollen is scattered by the insect

44

Plate 18: Half-open flower of Creeping Buttercup, (*Ranunculus repens*). (Enlarged).

world of the sun-lit air or by the wind, the plant, when flowering, enters a wider sphere to which, if we only relied on our senses, it would hardly seem to belong. Here contraction and refinement are but two sides of one and the same process.

With the formation of the stamens, the plant has in a way come to the end of a second stage. What follows now is in no way a direct continuation; for one cannot imagine the carpel as being a further metamorphosis of the petal-to-stamen process. There is no bridge here; a gap divides the two functions from one another. The essential part of the pistil (*gynaecium*) is the ovary. It contains the ovules and later ripens into the fruit. Only style and stigma have a certain similarity to stamens, and they also usually drop off at the same time. In the Buttercup, many ovaries are to be found close together in the centre of the flower. As no styles are formed, the stigmas are directly attached to them. Each carpel contains a single seed and develops into a dry, nut-like capsule which bursts only when germination starts.

Plate 19: Fruit of Field Buttercup. (Enlarged).

If one pursues the metamorphosis of the leaf destined to become an ovary, quite a different picture emerges from that of the metamorphosis of the stamen. There is no rolling in from the edges, but the whole leaf folds over and the edges grow together. It is of no significance whether, as in the case of the Buttercup, each carpel forms its own ovary, or whether several grow together into one (*syncarpous*). Goethe describes this in the above-mentioned book as follows:

Bearing these observations in mind, we cannot fail to recognize the leaf nature of the seed-vessel – notwithstanding their various forms, their special modification, and their relations among themselves. So, for

46

Consisting of one carpel.

Consisting of two carpels.

Consisting of three carpels.

Fig. 5: Diagrammatic drawings of the ovary.

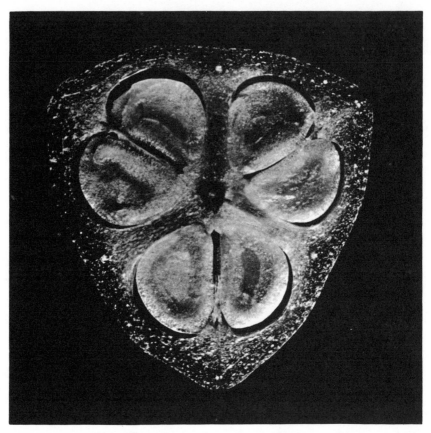

Plate 20: Cross section of a half-ripe seed capsule of a Tulip, showing three carpels. (Enlarged).

example, the legume would be a simple folded leaf concrescent by its margins, while siliquae would consist of several leaves, superimposed and fused. Compound seed-vessels would be explained as consisting of several leaves united round a middle point, their inner faces open towards one another, and their margins united. We may convince ourselves of this by observing the appearance presented when such aggregated capsules spring apart after ripening, since each member then reveals itself as an open pod or siliqua.

Goethe even mentions the case of carnations where the carpels are transformed back into sepals and, in place of the seed-capsule, a second flower grows out of the first – a sort of retrograde metamorphosis:

Plate 21: Dry Tulip seed vessels.

Retrograde metamorphosis, again, brings this law of Nature home to us. So, for example, in the pinks – flowers which are well known and loved for their very degeneracy – it may often be noticed that the seed capsules change back into calix-like leaves, and that the styles shorten correspondingly. Indeed pinks occur in which the seed-vessel has changed into a calix, real and complete, while its apical teeth still bear the delicate remains of the style and stigmas, and, from the interior of this second calix, a more or less perfect corolla is produced in place of the seeds.

One could easily fall into the error of thinking that the carpels too are 'contracted forms' and that they should be compared with stamens. One could even try to explain the formation of a carpel as a 'rolling in of the leaves', but the facts contradict this. During flowering the corolla and stamens are fully formed, while the ovary is but embryonic. Only during the ripening of the fruit is it shown in its

49

full expansion. This shows that it is no contracted organ, but, in many instances, surpasses all the other parts of the flower in substance.

When the ovary stands above the receptacle, it develops into a true fruit. In many cases, however, it is inside the receptacle so that the surrounding tissue becomes part of the fruit-forming process. The result of this is not a true fruit. For instance, the apple is not a true fruit, nor is the currant. Further examples will be mentioned later. The plant, which has expanded out into space during the dispersal of the pollen, turns back on itself. In scattering it gave itself to the surrounding world; now it withdraws into itself.

The seed is again a new beginning. For the third time the plant contracts to a minimum of material substance.

5   Flowers and Insects

A world full of wonder and mystery opens up when one studies the relationship between flowers and insects more closely. Two kingdoms of nature come together here and blend to become a third, an 'in between' kingdom. The insect which approaches the flower seems to lose something of its animality and acquires instead characteristics of the flowering world; the plant, on the other hand, seems to imitate the animal world – in appearance only, however, not in movement. Could one imagine a butterfly, if there were no flowers? Not only because it lives on nectar, but because, resembling a flower itself, its whole nature reflects the plant world. Conversely the flower also appears as a reflection of the butterfly, and so we feel a mutual interweaving.

Other insects also appear to cast their shadows over plants in order to appear as flowers. Popular names such as the Bee Orchid, Fly Orchid, Spider Orchid, etcetera, suggest yet another insect kingdom. Nowhere is there a more intimate, delicate reciprocity than between flowers and insects. 'They must be made for each other.'

The relationship between plant and insect is not limited to the flower alone. The entire plant and its stages of development have their parallel in the insect world. Let us take the butterfly as an example. It is a creature which seems as if it were spun from light. Its life elements are light and air, as water is for the fish. It touches the earth only fleetingly, fluttering from flower to flower. Its egg corresponds to the seed; it is entrusted to the earth yet does not become part of it. Just as the green shoot grows from the plant seed, so does the caterpillar emerge from the butterfly's egg. It feeds on leaves and lives on green plants. To grow is its main function, as it is the function of the shoot where node follows node and leaf follows leaf. As the growing plant repeats itself in this way, so the caterpillar consists of many similar segments. It moves in a rhythm of expanding and contracting. The plant in turn contracts at each node and expands in the succeeding internode.

When the time approaches for changing into a chrysalis, the caterpillar undergoes a great transformation, a real metamorphosis. It surrounds itself with a hard shell and dissolves. Not one individual part remains; all the organs are 'melted down' so that one cannot imagine a more complete transformation. Is it not very similar with the plant when it is about to form its flower? The bud is its 'pupa'. And how significant it is that the transformation from green leaf to flower is a jump! The plant has, in a way, to efface itself; its vegetative organs must be checked before a new higher stage can emerge. Thus the transformations in the two kingdoms resemble each other.

Few perhaps have the opportunity of watching a butterfly emerge from its chrysalis and seeing how it stretches itself until the folded-up wings are finally spread out. When a bud opens, the same thing happens. In the poppy, for example, it always seems quite miraculous when the petals emerge from their crumpled folds within the bud and become free and smooth.

The butterfly is not segmented in the same way as the caterpillar. We can differentiate head, thorax and abdomen. Legs and wings are attached to the thorax, antennae and mandibles to the head. The abdomen is without any attachments. So the rhythmic segmentation of the caterpillar is reorganized on a higher plane. The flower too, as we know, only becomes a complete organism when all the separate parts become concentrated in one point and become subordinate to the whole. This is what makes the flower appear animal-like: like does not follow like indefinitely, but each part appears as if guided to its ultimate purpose by the idea behind the whole entity.

Plant	*Butterfly*
Flower	Butterfly
Bud	Chrysalis
Shoot	Caterpillar
Seed	Egg

This unique concordance between the two kingdoms of nature, particularly the matching of flowers and appropriate insects, has been the subject of a great deal of thought and many theories. One does not, however, reach any satisfactory conclusion if one looks upon flowers and insects as separate beings each of which developed separately. The question of how the two have managed to adjust themselves to each

Plate 22: Leaf bud of Christmas Rose. (*Helleborus niger*). (Enlarged).

other subsequently will never be answered because it is not the right question to ask. It was Rudolf Steiner who from his scientific-spiritual research offered a solution which is convincing and makes sound common sense.

Rudolf Steiner describes how in previous ages of the earth plants and animals were a combined kingdom, a plant-animal or animal-plant kingdom. We have to imagine that there once lived creatures who combined the attributes of plants and animals. Indeed, plants today still have some things about them which seem to point to a common origin with the animals. There is for instance the variety of movements which occur in plants. In the flower particularly we find organs that are sensitive to touch, and even the green parts of plants often suggest that at one time they were more than just vegetatively alive. Creepers react to touch by specific twining movements. The well known Sensitive Plant (Mimosa) folds its leaves together as if it had been shocked. Many movements of leaves as they unfold suggest the fact that what now is a gesture was once imbued with feeling. If we could accept that flowers and insects once formed one of these 'in-between' kingdoms and that only after a certain time did they diverge from each other in their development, many questions would be answered. One part became plant by uniting itself with the earth, since when it grows towards its origin in the light. The other part, the butterfly, still retains its non-terrestrial, cosmic-nature, and has become a flower-animal. So it has happened that today flowers and insects belong together like two halves, positive and negative.

> Look at the plant –
> It is the earthbound butterfly.
> Look at the butterfly –
> It is the plant released by the Cosmos.
>
> *Rudolf Steiner*

We are confronted by a world which, in the truest sense of the word, has to be described poetically.

The flower, the mirror image, the negative of the insect, turns not only to the physical form of the insect but also joins in with the movements, the 'behaviour' of the visiting creatures. Accordingly, plants can be more or less specialized. The 'umbrella' of the umbelliferae, for instance, appears like a ready-laid table for everyone. Here we find all sorts of insects sucking the nectar from the juicy nectaries. Neither skill nor strength is required to obtain this precious food. At the other extreme are those flowers which are attuned to only one or two types of insects. So it has happened, for instance, that certain foreign plants grown in greenhouses were not able to produce seed because the insects belonging to them had not been imported with them. Between these two extremes are to be found every possible in-between stage. Flowers visited by butterflies are recognizable by having particularly long corolla tubes which exclude insects with short proboces. There is a great difference between flowers which are visited by butterflies and those visisted by moths. The butterfly *alights* on the flower. Besides the scent, it is also attracted by colour. Sometimes there are markings that lead it to the right part of the flower. The plant extends support to the insect's legs: it forms a 'landing stage'. Moths, by contrast, do not alight, but hover in front of the flower, merely inserting their long proboces into the corolla tube (e.g. Honeysuckle). The metamorphoses are manifold and wonderful and are influenced even by the different movements of insects.

Fig. 6.

Whether the anthers open on the inside or outside depends upon whether the insect approaches it from within or from without. Often there are mechanisms which function only if the right insect which fits the flower according to size, strength and behaviour, visits it (lilies, orchids, etcetera). Sometimes the flower tries to envelop the insect. It panders to the insect's peculiarities and even tries to cater for its feeding habits. The anthers and stigma may be in such a position that they brush the insect as it passes in flight. In some cases the flower manages to deceive the insect which can be kept prisoner for a while by blocking the exit with hairs. During its imprisonment the creature is fed, there being food-containing glands and hairs in the flowering prison. In other cases the insect has to struggle to achieve its purpose. If the insect is too weak it may perish or lose some of its limbs.

The multitude of adaptations is unlimited and one can easily be led to the wrong conclusion, that the plant has a reflective intelligence. However, the insect is the active part, the plant the passive one. The plant expresses in form what in the animal is activity. The animal's instincts become fixed mechanisms in the plant and that is why the insect fits the flower as a key the lock. Even if today they belong to two different kingdoms of nature in reality they are the two parts of a divided organism. Flower and insect are born together.

There is no need to quote examples of the flower-insect relationship, as many books have been written about it. One thing is clear: it is impossible to look upon the plant as an isolated object. Just as the plant, through its root, becomes part of the general life of the soil from

Fig. 7: Flower of *Gloriosa Superba* (Lily family), a woodland climber of tropical Africa. Imagine a tulip flower hanging upside down. Then let the six petals turn upwards. They again form a cup, but the inner side of the petals are turned outwards. The inside is now outside, and insects have to hover when they want to suck the nectar from the grooves at the base of the petals. To this movement the flower responds with a gracious bending of the stamens so that the six anthers are brought into such a position that the sucking insect must touch them. The plant senses the encircling insect. Even the forked stigma at the end of its long style is brought into the horizontal circle of the anthers. Such examples show convincingly that flower and insect together form an organic unity.

which it receives strength, so, when flowering, it is touched by the 'upper world'. This cosmic touch, filled with warmth and light, affects even the material functions of the plant and so removes it from the terrestrial realm, bringing a whole army of insects with it. Whether the adjustments are of a specialized or generalized nature is not so important. If we are not able to look upon this cosmic influence on the plant as something just as real as the life of the earth, then we will find it difficult to recognize the real world forces which are behind the entire being of the plant world.

6   *Leaf and Stem Tendencies as seen in the Flower*

Our studies have led us to realize that even the flower shows the effects of rhythmic contractions and expansions. The calix is the result of a contraction, the corolla of expansion. Contraction follows again as the stamens are formed (though their function – the scattering of pollen – is extension). In fruit formation – though a different principle is active – we again see expansion. The plant contracts a third time when the seed forms within the fruit. Tendency towards leaf and tendency towards stem are the two fundamental principles working in the formation of the plant. The stem tendency is vertical; Goethe calls it *geistiger Stab* – spiritual staff. It places the plant in the Earth–Sun alignment and gives it its stiff framework. The leaf tendency makes the plant expand horizontally. It makes it want to grow in a plane (two dimensional). This tendency furnishes the stem with leaves. The interplay of these two tendencies is responsible for the manifold forms of the plant world. Neither tendency could express itself without the other.

If the flower is a repetition of the plant on a higher level, it must be possible to find in it the same basic principles as in the green shoot. Needless to say, calix and corolla correspond to the leaf and, like the leaf, they spread themselves out. Even the stamens are born of the leaf tendency. They are a further development of the petals, and the transitional stages can still be traced. Corolla and stamens encircle the 'spiritual staff'.

It is a characteristic of the leaf tendency to express movement in the plant, while the stem tendency makes for rigidity. In the positioning of the leaves around the stem there is a kind of movement. If the stem itself wants to indulge in movement it can do so only when the leaf tendency gets the upper hand. Twining plants show this. They have no real stem and this is why they need support. What looks like stem is merely a connecting thread between the leaves – the expression of a spiral movement which has taken on substance, winding its way up from leaf to leaf.

Plate 23: Auricula
(*Primula auricula*).

The picture illustrates leaf tendency and stem tendency in the flower. At first the leaf tendency predominates in the flower, with the result that the flower stands out at right angles to the stem. When corolla and stamens have dropped off, the stem tendency asserts itself, and the pedicles carrying the ovary turn upwards. The seed ripens in the same position. (See picture).

Similar interesting changes occur in many flowers, even in the irregular though symmetrical Delphinium and Monkshood where, in contrast to the petals, the ovaries are regular and stellate but are dragged down into the horizontal by the flowers. When the flower fades, this influence is removed, and the ovaries turn upwards to their natural position. This is also to be seen in the Cowslip, Aquilegia, Crown Imperial, Lilium martagon, etc.

Drooping flowers turn upwards through 180 degrees after the petals fall. Thus we can see two quite different forces active in the flower; one in the petals, another in the ovary.

59

The expression of movement is much more frequently found in the leaf-like parts. Let us describe how a flower opens and closes. If it is to open, the inner surface of the petals must grow. Conversely, for the flower to close, the outer surface of the petals must stretch. This causes them to curve inwards. So it happens that for instance the petals of *Anemone Hepatica* double their length after having opened and closed several times. The same phenomenon can be observed in tulips and other flowers. The movement of the stamens follows a still different course. When the flower opens, we often find them bent inwards like little hooks, and they straighten up one by one as their anthers ripen. In some flowers (e.g., Delphiniums) one stamen after another bends towards the mouth of the flower to offer its pollen to the visiting insects. The stamens of plants like Berberis and Mahonia react at contact and fold inwards when touched.

There are some plants in which style and stigma, too, react and bend at contact but then style and stigma are related to the stamens as previously mentioned. They are often coloured, delicate and fragile like petals. They play no essential part in fruit development and sometimes drop off together with the stamens. In the Iris they resemble petals.

The ovary alone is that part of the blossom which expresses the stem tendency. When corolla and stamens are already fully developed, and style and stigma have attained maturity, the ovary is still in an embryonic state. The ovary is so placed that it is the direct extension of the flower stalk (peduncle). It is governed by the stem tendency. Quite often it has 'sunk into' the stalk; it is then called an inferior ovary. If there are more ovaries than one, they are arranged in a cylindrical spike along an extended stalk, pine-cone fashion. This is clearly to be seen in *Myosurus minimus* ('Mousetail), a member of the *Ranunculus* family.

In Goethe's *Sprüche in Prosa* (Prose Aphorisms) we find these words: 'It can be observed that the system of calix, corolla and stamens corresponds with that of the green leaves, whilst pistil, ovary and fruit belong to the same system as do the growth-buds. He who can see this clearly, is granted a deep insight into the secrets of Nature.'
Plate 24

Such sayings of Goethe must not be looked upon as the trivia of a genius. They are the outcome of years of conscientious observation. It is very satisfying and encouraging to find one's own thoughts

Plate 24: Foxglove
(*Digitalis purpurea*).

This deformation often occurs in the foxglove. The unusual form can be explained as follows: several flowers split and spread out and a new compound flower appears at the end of the stalk. The original flowers are now no more than petals. In the middle however several carpels grow together into a common seed vessel.

confirmed by the great naturalist. Goethe's 'growth-buds' and our 'stem tendency' are identical. If leaf and stem tendency are related to the whole Earth, one gets the following picture. Each plant stem, growing upwards, is like a reflected sunbeam. The leaf tendency, however, brings an element of enclosure; the totality of all the leaves around the Earth forms an enclosing, green skin.

Plate 25: Daisy, (*Bellis perennis*), proliferation. From the receptacle of the original flower-head sprout little daisies which even possess their own small green leaves. One can see that the daughter flowers form a unity, since they all face outwards to the circumference.

Our observations having led us to this point we shall be able to throw new light on the mysterious event of pollination and look at it from a hitherto unknown angle. Leaf tendency and stem tendency appear in the flower, the one appearing in the stamens, the other in

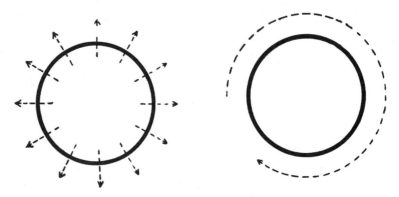

Fig. 8: Diagram of the stalk tendency and leaf tendency in relation to the whole earth.

the ovary. Pollination unites them. They fuse in the seed, endowing it with the power of germination. Concerning this we find another observation in Goethe's *Sprüche in Prosa*: 'In all living bodies we find the power to reproduce their own kind. When it appears parted in twain, we call it two sexes.'

Plate 26: Head of Dandelion in which the outer sepals have developed into green leaves. The flower is a repetition of the plant on a higher level. This is especially true of the *Compositae* to which the Dandelion belongs. In rare cases a rosette of leaves similar to those at ground level can develop in place of the sepals.

We found that it is a strict law of Nature that the ovary is the heart of the flower and can never be replaced by the stamens in the sequence of organs. Did we not know that the ovary is attached to the stem and is part of the vertical axis that continues right up into the flower, whilst the stamens are part of the leaf system, there could be no reason at all why there should not be flowers with stamens as their centre and ovaries arranged in a circle around them.

So our studies lead us to understand the inner necessity of outer facts and the laws which they obey.

Plate 27: Frost flowers on a window pane of a florist's shop. (Photo W. Firgan). An inexhaustible variety of forms can arise when water crystallizes in a thin film, as on a window pane. Plant-like forms appear, sometimes suggesting leaves, sometimes stems or even thistles. As our picture shows, forms can be grouped round a centre, their rhythmic shapes suggesting flowers.

Plants suggested by frost pictures are mainly those of the 'leaf type' such as sea weeds, palms, ferns, etc., but delicate moss-like forms are also not uncommon.

7   *Fruits*

When the calix and stamens have dropped off, the flower enters upon its second stage. It looks insignificant and requires a further period of development until the ovary has ripened into the fruit. Only after the flower has withered away does it start to grow and swell with nutritive material, or else it becomes a dried up pericarp (containing the seed).

The following observations are about fleshy fruits and the development of the plant will be discussed up to its final stages. Unfortunately it is not possible to study this final stage in the buttercup.

Let us consider a plant family whose characteristic is to produce as great a variety of fruits as the *Ranunculus* family does flowers – the *Rosaceae* – a family to which our fruit trees belong: apple, pear, quince,

Plate 28: Sloes.

peach, apricot, plum, cherry, as well as raspberry, blackberry and strawberry.

In contrast to the *Ranunculaceae*, the *Rosaceae* can only be understood as 'stem-plants'. Parts of the stem are metamorphosed into the many varieties of fruits.

Rose   Plum

Fig. 9.

Let us look at the cherry and the plum where the ovary itself becomes the fruit. Here the receptacle is bowl-shaped and hollow. Calix, corolla and stamens are attached to the edge, and the ovary, which is apocarpous (made of one carpel), rests free in this cavity. It swells up to become fruit when the flower has dropped off.

Fig. 10.

With the apple and the pear it is different. The ovary is composed of five carpel leaves and is inferior. Only the styles show between the stamens. The surrounding stem tissue becomes the fruit flesh, while the five carpels form the core containing the pips. That is why the apple is described as a false fruit (or swollen receptacle).

In between the apple and the cherry comes the rose. The receptacle is formed like a jug and becomes the fleshy rose hip; the ovaries are attached to the inner surface. In this case too the ovaries have no part in the formation of the flesh. They become the hard pips of the rose hips.

The construction of the strawberry is particularly interesting as in this case neither the ovary nor the stem tissue becomes the flesh, but the receptacle. As long as the strawberry is in flower, the many ovaries lie on the receptacle. As soon as it starts to grow and swell they are lifted up and spaced out. When it is ripe they are distributed evenly all over the fruit where they appear as little pips in the fruit flesh. The strawberry is not a true berry. It is really a swollen, aromatic receptacle.

Strawberry   Raspberry

Fig. 11.

Raspberry and blackberry represent another type of fruit. Here the individual ovaries develop independently into fleshy little fruitlets. What is called a berry is in this case a compound fruit as it is composed of many equal parts. The calix and a cone-shaped core remain attached to the stem when the berry is picked.

If one looks at all five examples together, one realizes that the fruit principle has a vertical tendency. The leaf realm from which the metamorphosed corolla originates is not touched at all. The fruit is the

67

climax of the stem principle of the plant. All the fruits develop from the ovary or stem tissue. Malformation can be seen in this connection. For instance it has been known for cucumbers to grow leaves from the elongated fruit. This would not be possible if the cucumber were not a metamorphosed piece of stem. It has developed from an inferior ovary.

When the plant reaches the fruit stage it opens out a second time towards its surroundings. Once again qualities are developed which cannot be understood as coming from the plant alone. A second time the plant as it were over-reaches itself, produces colour and scent and, as during pollination, develops a connection with other creatures, this time the ones which eat its fruits. So in the blossom we have two separate phases, the second one starting when the first one is finished. Flowering and fruiting are by no means the same thing. The fact that the ovary is green and the corolla and anthers are coloured should teach us that here we deal with two quite different formations one of which is related to the leaf and the other to the stem. What the bud is to the flower, the ovary is to the fruit. Correspondingly, the metamorphoses are different during the two different phases. Fruit formation is a process which takes place mainly in material substances, so that we cannot speak about metamorphosis in the same sense as we do regarding the corolla. In the autumn, when the fruits are ripened, we see the material result of the co-operation of plant and cosmos 'dripping' from the trees and bushes. The drop is the archetypal form of the fruit which appears pure in the berry, elongated etcetera in other types. Segments, notches and grooves indicate the original number of carpels.

A threefold rhythm is the basis of plant growth. It grows from the seed where it has been compressed into the minimum of space. In the green leaves the first expansion takes place. This is of a purely vegetative nature, but from it arises the next compressed phase in the bud. The second expansion occurs in the corolla. But in the heart of this leafy structure we have another 'seed' stage, the ovary. This swells into the fruit, and here we have the third expansion. At the same time there occurs an internal, third contraction when the plant 'slips into' the compressed space of the seed. The circle has been completed.

Twice the leafy element and only once the stem tendency develop their highest qualities. In between occurs pollination. The two basic forces of plant formation are united in a wonderful synthesis. A new whole has been created capable of development.

The usual reason given for the fruiting of plants is that as animals eat the fruit they distribute the seeds at the same time. The inadequacy of such an explanation is usually overlooked. After all, plants with non-succulent fruits achieve the same end by different means. Nature is most inventive in its methods of seed distribution. The formation of succulent fruits is only one of these and is, from the point of view of the plant, most unecomomic. What a variety of substances are required to produce a ripened fruit! Rather, nature has destined the fruits for the stomachs of hungry animals. The plant could do without the fruit. They are given as an almost useless (to the plant) excretory product to the hungry environment. However, plant and animal kingdoms form a living whole in which one cannot be thought of without the other.

The fruit is the last and highest stage of the complete plant-forming process. If this stage is missing the process has come to an end too soon and we must ask for the reasons. Examples will be given later. The complete plant-forming process requires that the transformation of substances reach its highest point in the fruit. All the stages are included, from the salty nature of the root via the watery nature of the green shoot to the aroma of the fruit. Thus every kingdom reflects the totality of the whole of nature. Later it will be discussed how, in the plant-forming process, the threefoldness of the human, animal and earth organisms are contained. Then it will be understood what the process of fruiting in the plant kingdom really means.

8　Trees

Trees, without doubt, are one of the most remarkable representatives of the vegetable kingdom. Their expressiveness fills the student with admiration. The 'gestures' expressing the formative forces working in them are not to be found in the vegetation growing actually on the surface of the earth. The soil itself helps to give form. Its living force rises above its normal level and carries the vegetative life with it.

Rudolf Steiner, who gave us many interesting indications regarding trees, spoke of their trunks as protuberances of the soil. This idea seems very unusual but proves extremely helpful for a deeper understanding. When we speak of soil, we must not however think only of its dead, mineral part. In one of his lectures Rudolf Steiner says: 'The soil, in addition to the fact that it contains minerals, has within it forces which burst forth into something plantlike. This thrusts itself up and becomes the tree trunk. And that which grows on the trunk can be likened to the herbaceous plants which grow directly out of the soil.' Where trees grow, plants are lifted up from the soil and spread out on the branches. The coming-into-leaf of a tree is the same process as the germination of a plant from seed or its sprouting from a root stock. Reversing the picture we can say: The soil is the trunk of herbaceous plants.

Certainly one could object at this point, knowing that a tree trunk consists mainly of cellulose (wood) which was formed by life processes in the green parts in the presence of light and then built up. How, therefore, could a tree be 'raised up' soil? And yet it is the power of the earth which leads to the consolidation of the tree substance. It is the earth which gives it its lasting quality, consolidating carbohydrates into wood. The tree trunk is soil become plant substance. If we find an ant hill, we recognize that it was built up by the ants, although the mound consists of pine needles, little stones etcetera, not of ants.

There is, however, a considerable difference between herbaceous plants and shoots on trees. Herbaceous plants root in the mineral earth. Therefore they have to pass through all the stages from coty-

Fig. 12.

ledons via the simple leaf to the fully developed one, until they can finally produce the flower. Trees on the other hand need not start the plant-growing process at its first beginning. They are able straight away to grow large, fully developed leaves; in fact in many instances flowers appear without being preceded by any green parts. In no time the tree presents itself in its full leaf-and-flower glory. The speed with which it is done is self-explanatory. The young shoots grow out of something – the trunk – which is itself semi-plantlike. 'Trees seem the noblest of plants because their countless leaves and flowers depend only indirectly on the soil and are already almost plants upon plants' (Novalis: *Fragments*).

Although we are right in considering the wood to be a substance which, though on a plant level, corresponds to the dead mineral rock, we must not forget that it is half-alive. Only the innermost core, the heart-wood, has become completely lifeless and serves solely as support. The peripheral layers, the sap-wood, is alive and performs important functions regarding the conducting of water and the storage of organic substances such as starch or protein. If an old tree becomes hollow (limes, willows), the heart-wood decays and only the cylindrical part of the trunk, the sap-wood which surrounds the heart-wood, remains alive.

Very active life takes place in the bark. The dead parts are cast off either in flakes (plane tree), rings (birch), scales (oak), or long strips as in vines. So the trunk increases in girth, adding layer after layer of wood and growing a new bark all the time, only to cast it off when it has reached a certain age and has become too tight for the tree. Rejuvenation of the tree takes place in the cambium, that delicate, glutinous layer which is situated between bark and wood. It encases the wood from the root of the tree to the tips of its twigs like a tube. In spring, when the sap rises, it is very easy to peel off the bark and cambium from the wood.

The life of the cambium connects all parts of the tree. For the shoot it takes the place of the roots. But to understand this comparison, roots must be looked at as they have been studied in previous chapters. The inner *function* is what matters, not the outer appearance. Like the finest root tips the cambium is ceaselessly active and productive. The comparison holds good especially where the cambium borders on the wood, for in both cases we see the hardening processes setting in very soon. We described this characteristic of the root which is intensively alive but soon gives way to hardening. The layer of wood produced by the cambium within a growth period is called an annual ring. The number of annual rings tells us the age of a tree; their thickness, whether the year was a favourable one or not.

Let us picture the great difference in the development of growing-point and cambium. The cells of the growing-point form stalks, leaves and flowers after going through manifold transformations. The cells of the cambium, on the contrary, take the shortest route towards an early hardening process by producing wood and bark *without ever having been a plant in the real sense of the word*. They take no part in the actual plant formation but supply the bulk for the building up of the wood.

Plate 29: Date Palm, Cypress and Bay in southern Dalmatia.

Seen from this point of view, every tree shows two opposing tendencies. In the first, the cosmic one, formative forces are at work which make it an image of the extra-terrestrial, and a stranger on the earth. The other shows the hardening mineral force of the earth. Though bark and wood are not identical with the mineral part of the earth's crust, they are on the way towards it. A tree trunk is a vegetative rock, and lichens and mosses grow on it as they do on the barren ground of the mountains and the arctic tundra.

So we see what a true image Rudolf Steiner has set before us. The terrestrial force truly thrusts upward when building tree trunks. The cambium is the organ which helps it to draw the mineralizing tendency upwards above the earth. Where tree trunks grow, the earth is still in the process of becoming. The earth's crust is the end product of a life process, and it is not a playful comparison to liken the annual rings to geological layers, though on the higher level of living substance.

All kinds of insects have their home in the mould of hollow and rotting trunks, just as other insects live in the soil. Their larvae develop there. It is incorrect to call them parasites; rather, they have chosen for the scene of their development a level one step higher, i.e. from soil to 'thrust up soil', which apparently is better suited to their needs.

The earthiness of the wood as compared with non-woody parts of the plant also shows in the greater ash content of the former. Ashes are substances so earth-bound that they neither evaporate nor burn up (saltlike).

The 'thrust-up soil' however, lifts to a higher level other things besides plants. The picture can change from the tranquillity of a meadow to the dramatic effect of the primaeval forest. Here the new level is used by creepers, lianas, etcetera, as well as birds, reptiles and even mammals. A second and different sphere of life has been created above the first.

A tree has a 'within' and a 'without' just as much as it has a 'below' and an 'above'. Like the earth it is on the way to becoming an organism. So it is clear that compared with the other plants, man experiences a special relationship to trees, which is evident from their particular significance in mythology, folk lore and old customs. Trees were always felt to be an expression, indeed a dwelling place, of unique spirituality, and as such were revered, worshipped or even feared.

Finally, the difference between trees and herbaceous plants must be viewed from still another angle. One can easily make the mistake of thinking that tree trunks are simply the continuation of green stalks, the crown only an inflorescence upon which duration is bestowed. However, this is not so. To grasp the difference rightly, consider the development of our fruit trees. In the case of annual, non-woody growth the stalk is the 'staff' which carries the plant into the light. As in all the *Rosaceae*, the phylotaxis of the leaves follows a two-five rhythm – (see Chapter 2, *Germination and Propagation*) – and this placing of the leaves at first also influences the branching. As soon, however, as the tree begins to produce wood and adds its crown, it is influenced and transformed by formative forces of an entirely different nature. We recognize the branch systems of individual trees even if they are bare of leaves. These systems reveal individuality, and it cannot be sufficiently emphasized that the types of trees cannot be explained from the placing of their leaves. Certainly the process of wood production makes use of the upward-striving stalk but in its nature there is something else, indeed something opposed to it. It makes, so to speak, an entirely new organism out of the former vegetative form. In wood formation the formative forces appear which make an apple an apple, a pear a pear, a fir a fir tree. These it is which produce the extended umbrella of the pine or the contracted pyramid of the Lombardy Poplar; they set before us contrasts such as the oak and the birch and make each tree an individual.

Of course, this is not to say that the formative forces lie in the wood substance but only that the particular essence of a tree appears first at that moment when the plant structure is hardened into wood.

It is a most rewarding study for the botanist to follow up how a rose sapling becomes a shrub, from a mere upward-striving, vegetative shoot. Shrubs are half-trees, only crowns.

Rudolf Steiner's idea of a tree trunk as thrust-up earth forms a living image which can help us to become aware of a secret of nature of the greatest importance which is usually passed over, since trees are something so obviously familiar.

9   Goethe's Archetypal Plant

Goethe's idea of the archetypal plant and its scientific significance has till now been very little understood. It would probably not have been noticed at all, had not Rudolf Steiner spoken and pointed out that Goethe had been a pioneer in botany just as in many other spheres. That his work was but little appreciated and usually very much misunderstood is beside the point.

Rudolf Steiner points to Goethe as the Copernicus and Kepler of the organic world. He is the first person who clearly described where Goethe's great service to science really lay. It is not in the single observation that led to the idea, but in his method of observing the plant kingdom that the basic principle of the archetypal plant is to be found.

> It is not a question of emphasizing the fact that leaf, calix, corolla etcetera are plant organs identical with each other and unfolding out of a common basic form. The essential point is Goethe's conception of the whole plant-nature as a living thing, and how he thought of the individual parts as proceeding from the whole. His *idea of the nature of the organism* is his central, most individual discovery in the realm of biology.*

The archetypal plant as Goethe saw it is something fundamentally different from the ancestral types or primitive forms postulated by natural science today. It is not a physical nor even a theoretical progenitor of descendants which could be derived from it according to the modern theory of descent. It is a principle which – as Goethe expresses it – is laid into every plant on earth as a common model.

On April 17th, 1787 in Palermo, Goethe said of the archetypal plant: 'It must needs exist, for how would I recognize a plant as a plant if they were not all made after the same pattern?'

The way Goethe came upon his idea is characteristic. The question of what it is that makes a plant a plant, what all plants of the whole

* Rudolf Steiner. *Goethe's World Conception.* Anthroposophical Publishing Company, London, 1928.

world have in common, was of the livliest interest to him. He wanted to know what 'planthood' really was. In the many different forms of plants Goethe saw the revelation of one archetypal being spread out in great diversity. The archetypal plant had to be thousandfold in the world of phenomena but uniform in its innermost nature. In a letter to Herder we read:

> I would have you know that I have come very near to the secret of plant creation and that it is the simplest thing you can think of. The archetypal plant is to be the most wondrous creature of the world and Nature herself will envy me it. With this model and the necessary key one can invent plants without number, consistent plants, i.e. plants which, although they do not exist, could exist and are not mere poetical shadows but possess an inner truth and necessity. This law could be expanded to cover all living things.

The archetypal plant is leaf, for the whole plant develops from the leaf. On May 17th, 1787, Goethe writes to Herder:

> It has come to me that in the organ which we are wont to call the leaf lies hidden the true Proteus who could hide or reveal himself in any form. From beginning to end the plant is but leaf so inseparably united with the future growing shoot that one cannot picture one without the other.

Goethe got to know his archetypal plant not by investigating primitive and less perfect vegetation as is usually done today, but by studying the higher plants which take their archetypal element, the node and its adjoining leaf, step by step up the 'spiritual ladder' so classically described in his poem *The Metamorphosis of Plants*:

> None resembleth another, yet all their forms have a likeness;
> Therefore a mystical law is by the chorus proclaim'd;
> Closely observe how the plant, by little and little progressing,
> Step by step guided on, changeth to blossom and fruit!
> First from the seed it unravels itself, as soon as the silent
> Fruit-bearing womb of the earth kindly allows its escape,
> And to the charms of light, the holy, the ever-in-motion,
> Trusteth the delicate leaves, feebly beginning to shoot.
> Leaf and root, and bud, still void of colour, and shapeless;
> Upward then strives it to swell, in gentle moisture confiding;
> Yet still simple remaineth its figure, when first it appeareth;
> Soon a shoot, succeeding it, riseth on high, and reneweth,
> Piling up node upon node, ever the primitive form;
> Yet not ever alike: for the following leaf, as thou seest,
> Ever produceth itself, fashion'd in manifold ways.
> Longer, more indented, in points and in parts more divided –
> Yet here Nature restraineth, with powerful hands, the formation,

And to a perfecter end, guideth with softness its growth.
So that the figure ere long gentler effects doth disclose,
Soon and in silence is check'd the growth of the vigorous branches,
And the rib of the stalk fuller becometh in form.
Ranged in a circle, in numbers that now are small, and now countless,
Gather the smaller sized leaves, close by the side of their like.
Round the axis compress'd the sheltering calix unfoldeth,
And, as the perfectest type, brilliant-hued coronals form.
Yes, the leaf with its hues feeleth the hand all divine,
And on a sudden contracteth itself; the tenderest figures
Twofold as yet, hasten on, destined to blend into one.
Lovingly now the beauteous pairs are standing together,
Gather'd in countless array, there where the altar is raised.
Presently, parcell'd out, unnumber'd germs are seen swelling,
Sweetly conceal'd in the womb, where is made the fruit.
Here doth Nature close the ring of her forces eternal,
Yet doth a new one, at once, cling to the one gone before.

(shortened version)*

Later on Goethe called the archetypal plant 'sensible-supersensible'. Supersensible in its nature, it nevertheless can be beheld with the very eye when it reveals itself, now in the cotyledons, now in the more or less developed leaf or in the different organs of the flower. It is alive, which means it changes continuously, and yet always remains the same.

The archetypal plant is not the threefold being of root, leaf and flower which Rudolf Steiner outlined as being the reverse picture of threefold man. It is merely the leaf at the node of the stem, and, in the realm of ideas, a key to understanding the metamorphosis of plants, and therefore fundamental for the comprehension of animate nature. It is in no way related to man.

The difference between Goethe's archetypal plant and the plant picture outlined by Rudolf Steiner is apparent the moment we look at the root. This is an independent member of the plant and, according to Rudolf Steiner, corresponds to the human head. In Goethe's conception of the plant the root plays a subordinate part as it cannot be derived from the leaf.

Rudolf Steiner, whose ideas give the background to the following chapter, understood the metamorphosis of the plant in a much wider sense. What Goethe developed from his archetypal plant he applies to the whole plant and thereby connects it to man as well as to the whole earth.

* From the translation by Edgar A. Bowring. George Bell and Sons, London, 1874.

II   The Living Face of the Earth

10   From Pole to Equator

The metamorphosis of vegetation from the poles towards the equator falls into place when the whole earth is understood as a living organism. Its life is determined by the interplay of forces from within and without. The plant is the organ which stands between these forces. The interplay of forces finds an immediate expression and becomes visible in the plant forms, and he who can read the message looks straight into the grand features of the Earth's face.

Even a superficial glance at the path of sun and stars can make us realize what great differences there must be between the two extremes. In the temperate zones the sun looks down on us obliquely. According to the time of year its arc is larger or smaller, but it never rises to the zenith. Looking southwards we see the fixed stars and planets rise and set as sun and moon do. In the northern sky those stars circling the pole star never set.

At the equator the sun rises vertically into the sky and sets vertically. Its arc is never oblique, but larger or smaller according to whether it is in the central position over the equator, or shifts to the south or north. At the time of the equinoxes it is largest; the sun passes the zenith and at midday a vertical stick throws no shadow. In the tropics the axis around which the vault of heaven turns runs south-north. The pole star lies on the horizon.

Needless to say, the sun's influence must be more intense in the tropics. Its rays come down more steeply; they are able to penetrate deeper and their influence lasts longer. Both light and heat are responsible for the abundance of tropical vegetation, and even the greater number of stars has an effect. All the stars of heaven appear above the horizon every day while in Europe, for example, we see only the stars of the northern heavens; others never become visible.

The polar regions are a complete contrast to the tropics. In the tropics every day is like a brief summer. At the poles the sun rises only once a year, in spring. The year becomes a day. Without setting, the sun climbs higher very gradually but circles rather close to the horizon. It is highest at midsummer, which is the midday of the polar day. Then it circles lower and eventually sets, the polar night lasting half a year. Moon and stars, too, move in parallel circles over the horizon. At the North Pole, the pole star, the centre of the vault of heaven, stands in the zenith.

The fact that the sun shines for only half a year and then without interruption and with its rays striking the earth at an oblique angle is of great significance. Nor can it be without effect that the stars looking down on the polar landscape never change. One half of the firmament does not exist for the polar region. Stars neither rise nor set.

This monotony finds its expression in the polar flora. Compared with the superabundance of species in the tropics and the temperate zones, the polar regions show but a limited number.

If one could see the whole earth from above, it would look like a living map. Regions like tropical South America, some parts of Asia and Africa would show the superabundant growth of primaeval forests. The temperate zones would be characterized by mixed forests and green fields; areas like deserts and steppes would appear more or less bare of plant growth. Near the pole, however, beyond the tree limit, the arctic regions with their sparse growth would display an unusual sight. Like no other region, the arctic shows us to what extent plant growth depends upon extra-terrestrial influences. Water, the life blood of the earth, is frozen for the greater part of the year. Rocklike, it covers large areas of the land and even where it is not solid, it lacks the power to induce life. Dead are the waters of the arctic.

Vegetation awakens in stages from the extreme north southwards. It wrests itself free from the ground step by step, and the first distinct plant growth is known as the tundra-formation. Even so, the tundra is a picture of desolation. No tree, no shrub raises a leafy crown; all plants are as if welded to the ground and unable to lift themselves up. The characteristic plants are mosses and lichens. In the North American tundra the lichens predominate; in the northern regions of the Asiatic continent, the mosses.

Travellers tell us that the vegetation of the tundras often resembles just a verdigris covering over the naked earth. Certainly the concept 'vegetation', bringing to mind the leafy forests of the more southern zones, does not rightly apply to this landscape. All arctic plants are

Plate 30: Limit of the conifer zone in Russian Lapland. (From Schimper, *Pflanzen-geographie auf physiologischer Grundlage*).

contracted to dwarfedness, hardened in all their parts and finely formed. Many species form cushions and low mounds, often hemispherical in shape. Though they are higher plants, outer circumstances give them moss-like and lichen-like shapes.

Only south of the forest limit does life awaken sufficiently for timber growth to become possible. Certainly the first tree specimens at the forest limit are not to be compared with the trees of the temperate zone, let alone with the giants of the tropics. Isolated trees with frost-damaged branches give the impression of roots sticking up into the air. Of special interest are trees like the spruce which pushes its way north beyond the forest boundary into the tundra. What does this tree look like under arctic conditions? It loses the power to stretch upwards and remains welded to the earth. The stem branches out directly over the surface of the soil and spreads in the form of concentric prostrate branches. At the circumference the tips turn up and give the impression of a small thicket of dwarfed shrubs or many individual tiny trees. Only a close examination shows the inner connection of the whole. One sees that the arctic conditions compel the stem to cling to the earth like a sort of root.

A similar transformation occurs in the arctic willows. 'Baer describes three species of willows in Novaya Zemlya. The shoots of the woolly willow (*Salix lanata*) are 5 to 6 inches high, another rises to 4 or 5 inches above the ground, and the smallest (*Salix polaris*) is only half an inch high. The latter develops only two leaves with a single catkin but, like the others, a mass of these tiny plantlets is borne on a dense wide-spreading branch system united into an individual whole by a subterranean vertical stem' (from Grisebach *Vegetation der Erde*).

Plate 31: Shrubby conifer at the extremity of the tree line in Russian Lapland. In the foreground is a single bush which branches out at ground level. The tips of the branches stand up like minute trees and give the plant the appearance of a thicket. (From Schimper, *Pflanzengeographie auf physiologischer Grundlage*).

Such examples could be multiplied. They would all show the same thing: that the plants of the polar regions are formed by hardening and contracting forces which give to the entire plant characteristics which are normally concentrated in the root. Luxuriant vegetative growth can never develop. These tendencies help to explain the conifers which are characteristic only of the northern hemisphere. In place of leafy foliage are hard needle-leaves; contracting tendencies let nothing remain but a sort of midrib. All growth is as if immobilized in the hard wood.

It would, however, be a mistake to conclude from our former remarks that the arctic vegetation is flowerless. On the contrary, the very short summer charms out of the depressed cushions and mats an often extravagant superabundance of flowers, the colours of which are intense and incomparably pure. The traveller stands before it in wonder. The masses of flowers shimmer like a magic veil – an exhalation – over the stunted vegetation. In places exposed to the sun, seeds germinate when the snow has melted and develop very fast, flowers shoot up, nodding on their stalks, and insects fly amongst them. Soon the seeds ripen and with that the whole fairyland disappears, only to resume its play the following year.

Opposites seem to meet in the arctic without transition: summer and winter, light and darkness, heat and frost, earth and air. In the plants, root-tendency and flower-tendency are juxtaposed but without interpenetrating. The arctic flowering plants are divided into their polarities: hardened and contracted shoot and star-like flowers. The true plant, i.e. the vegetative shoot, has been left out. Extra-terrestrial influences – light, warmth etcetera – are reflected by the earth in the polar regions, whereas in the tropics they are absorbed. The Earth 'shines' in the arctic, as Rudolf Steiner puts it, and one of the ways in which it shines is expressed in the splendour of its flowers, not darkened by earth conditions but like a reflection of cosmic forces.

Leafy growth develops farther south. In the inner parts of Asia, for instance, we find those magnificent mixed forests composed of deciduous and coniferous trees. Deciduous trees come to their fullest development in the tropics where the Earth is tremendously enlivened by the sun's influence. The tropical rain forest is the culmination of this development; there is not enough space to hold the mass of plants. The gigantic trees are the foundation of the tropical jungle,

Plate 32: *Ficus magnoloides*. This enormous fig tree sends out adventitious roots vertically from its branches to the ground, which in the course of time thicken into trunks. In this way a single tree becomes a little wood.

their immense crowns crowding the air and intertwining their branches. Only the strongest can survive and when one giant falls it takes in its fall all plants around it and under it. Each tree is like a garden covered by other plants from top to bottom; nowhere does its bark show and climbers and creepers grow on it in untold variety. Even the space between trunk and branches is filled with lianas.

When we hear the lovely word Liana a whole series of images wells up from the dim recesses of the mind, reminiscences from early years in sharp outlines and bright colours. A leafy vault spreads above the gigantic treetrunks rising upwards like the columns of an immense hall. Here and there a thin shaft of sunlight penetrates it. On the forest floor, fallen trunks are covered by the rank growth of ferns and bushes. Brown, twisted roots prevent any further advance into the still dusk. In contrast with the creepy depth, how colourful the glades at the edge of the forest! Wildly entangled plants of all kinds form an impenetrable thicket, grow higher and higher up to the crowns of the giant trees, and shut off from view the interior of the forest. This is the true home of the lianas. Everything rambles, winds and climbs on everything else, and we try in vain to find which stems and leaves and flowers and fruits belong together. Here the lianas weave and work green walls and tapestries, there they hang in swinging garlands or drop from branches in wide curtains. In other places they span gaps between branches or trees in lush coils, build bridges, leafy passages and arcs. Isolated trees become green pillars or more frequently supports for green pyramids with the crown spread over them like an umbrella. When the lianas and their supports have become old and their winding stems leafless, they look like cables stretched between ground and crown. Sometimes taut, sometimes slack, they rise up from the ground and lose themselves above. Some are twisted into thick cables, others hang like corkscrews or again are flattened into bands or form stairs.

High up the garlands and wreaths of lianas bear the most colourful flowers. A bunch of fiery sparks shows here, a long blue grape cluster hangs there in the sun, and elsewhere a dark wall is embroidered with hundreds of blue, red and yellow flowers; and where there are flowers, and fruits ripen, their guests are never far off: the multicoloured butterflies and the birds. Their playground is the liana-clad edge of the forest. (Kerner-Hansen, *Pflazenleben*).

It would be hard to describe the tropical forest in more colourful pictures or better words.

Finally there are the Epiphytes which constitute an important part of the tropical forests. These are the plants which do not root in the soil but live higher up in the crowns, on branches and in the hollows of

Plate 33: Travellers Tree (*Ravenala madagascariensis*), one of the Banana family which, in spite of its size, has reached only the status of a herbaceous plant with a soft stem. (Drawn from nature in Kerner's *Pflanzenleben*).

tree trunks. Dew and rain nourish them and rotting leaves and other humus matter is their soil. They consist mainly of a great variety of colourful orchids and some ferns. Many of the tropical plants seem far too large for their type. The African grasses can hide a man on horseback. The Banana and its relative, the Travellers Tree of Madagascar, for instance, in spite of their great size, are structurally on the level of herbs. If reduced in size they might well be found among meadow plants.

So heat as opposed to frost reveals its power in the plant kingdom. It makes everything expand, and gigantic formations appear. Vitality too is enhanced. The stem of the Indian Fig Tree, even when very old, is able to join its own stems together into one, like a graft.

The temperate zones are transitional between north and south. They are the fulcrum of vegetation. One side of the scale is weighed down to

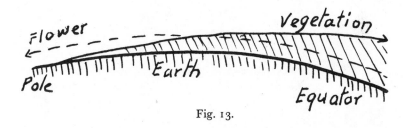

Fig. 13.

the ground at the poles, the other goes up at the equator. In arctic vegetation, root and flower are as it were divided. In tropical plants everything is intermixed. The flowering impulse works down into the leaves and stems, even into trunks and roots, hence their aromatic qualities. Where bark, wood or leaf becomes a spice, processes normally working in flower or fruit find their way down to other parts. The flowers and fruits themselves have more inner heat and fire, sometimes give off intoxicating or narcotic scents, and even have weird demonic forms. The flowering impulse is so strong in some plants that the flower predominates, as in orchids. The plant is almost only flower. The epiphytes are flowers taken off the ground and tossed up into the air as if on a high tide of vegetation – emancipated blossoms. Epiphytic ferns arrange their leaves funnel-fashion, thus copying the calix of a flower. *Bromeliaceae* (pineapple) also arrange their leaves like a funnel and have, in addition, coloured leaves.

87

Plate 34: Dragon's Blood Tree (*Dracaena draco*). Southern tree from Teneriffe with primitive branch system, belonging to the Lily family. Clusters of red berries hang from the leafy crown. If one scores the bark with a knife it exudes a thick red juice, the 'dragon's blood'. (Author's photograph from the botanical gardens in Palermo).

Plate 35: *Aloe arborescens* (Tree Aloe). This African plant (height 4-5 metres) is also of a primitive type. Plant families which are only herbaceous plants in the temperate zones, become trees in the tropics. This is an expression of the strong earth forces which in the tropics are enhanced by the life-giving power of the Sun. (Photographed in the botanical gardens in Palermo).

The pure and clear cosmic nature of the polar regions is in contrast to the extreme earthly nature of the tropics.

EARTH

Polar regions	*Tropics*
Frost – ground dead	Heat – soil living
Plants contracted	Plants expanded
Growth closely pressed to the ground	Growth lifted up into the air
Flower distinct from plant	Flowering impulse submerged in plant.
Forming Forces	Distending Forces

Though this is only a diagram, it helps us to penetrate the profoundest laws we know, those of the Earth organism. The great body of the Earth has its polarities just as the human body has.

We may well ask what is the meaning of these polarities. The next chapter will offer some enlightenment. However, it will first be necessary to describe the threefold division of the plant into root, leaf and

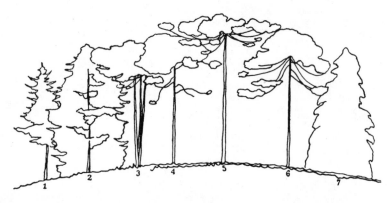

Fig. 14: Metamorphosis of the conifers from pole to equator (From Usteri). 1. *Larix* (larch), up to 70° North. 2. *Cedrus atlantica* about 32° North. 3. *Pinus (Canariensis?)* about 30° North. 4. *Pinus Hardwegii*, Popocatepetl, about 20° North. 5. *Araucaria Brasiliana*, northern limit about 29°. 6. *Araucaria imbricata* (Monkey puzzle), southern limit about 40° S. 7. *Libocedrus Chilensis*, to about 44° S. and even more. Tree forms resembling inflorescences with umbrella-like crowns stand in the middle, flanked by more conical forms towards the poles.

flower in connection with the threefold nature of man. It will then become clear that the macrocosmic organism Earth can really be understood only when seen in connection with the microcosmic organism, man.

The plant finds its place between this macrocosm and microcosm. It mirrors the laws inherent in both. Whether root-tendency or leaf-tendency or flower-tendency prevails depends, among other factors, on the region of the earth in which the plant grows. Through this ability to become one-sided, vegetation becomes a gigantic mirror, in which images clearly reveal (to him who understands how to read them) the laws of the Earth organism. Some variations of these basic laws will be described in the next chapter.

11 From the Lowlands to the Eternal Snows

Every mountain repeats in miniature what the Earth as a prototype is as a whole. Every snow-capped summit is like a little pole with eternal snow. The strong impression one has when descending from a snow-capped mountain has often been described; within a few hours there passes before one's eyes what would be experienced only in weeks on a journey from pole to equator. Alexander von Humboldt wrote in *Kosmos*:

> The mountainous region near the equator has one advantage not sufficiently observed. It is that part of our planet where, in the most restricted space, the variety of impressions reaches its maximum. In the deeply furrowed range of the Andes in New Granada and Quito, one may see all forms of plants and all the stars of heaven at the same time. One glance encompasses the Southern Cross, the Magellan Clouds and the pointers of the Bear. Here the womb of the Earth and both hemispheres of the heavens are open and display their riches and manifold forms. The climates, and the different zones of plant growth determined by them, are laid one above the other in strata. The laws of decreasing warmth are made clear to the observer, and graven with eternal strokes on the rocky walls of the slopes of the Andes.

The Earth is no abstract globe but a living body. One can study the basic laws of life, but one should not expect these to reveal themselves simply. We rather have to do with the interweaving and working of polarities, which is not a simple matter of physical laws, but a true life process. The metamorphosis of the vegetation from the high mountains to the lowlands is comparable to that from pole to equator. Alpine plants show the same characteristics as arctic ones, brought about by the same earthly conditions. Sometimes we even find the same plant species both high in the mountains and in arctic countries. The arctic willows closely resemble those recumbent and spreading willows of the mountains which illustrate the transition from tree to shrub, and finally, to the low cushion vegetation.

Plate 36: *Soldanella alpina* breaking through the snow.

For the mountains are like cold islands in the sea of the warm and temperate plains; they are fragments of the poles and the way from the plain up into the arctic of the high peaks takes the mountaineer through an instructive compendium of the different zones. A walk on the southern slopes of the Alps is especially exciting: We turn away from the Lago Maggiore (1965 metres above sea level), the olive trees, cypresses, orange and pomegranate trees of the Mediterranean climes and their chestnut groves, and walk in the shade of central European beech woods; then we inhale the aromatic fragrance of the Nordic spruce and larch forests, enjoy the beauty of the flower-studded alpine meadows above the tree line and find ourselves surrounded by a great number of plants of arctic type, right up to the summit of Pizzo Centrale (3003 metres). Here we stand among plants from Spitzbergen and Greenland, and look down into the southern valleys. So it happens that 2800 metres difference in altitude is equal to a journey from the 40th to the 80th parallel, a distance of

Plate 37: *Dryas octopetala*, an alpine shrub belonging to the *Rosaceae*. The branches form a network over the ground. The shrub is evergreen and has leathery leaves, furry and silver on the underside.

about 9000 kilometres, while the distance from Locarno on the Lago Maggiore to the Pizzo Centrale is 48 kilometres, as the crow flies (Prof. C. Schroeter: *Alpenblumen*).

Mountains in the tropics illustrate this particularly well as they present plant pictures of all the zones of the earth, from the extreme north to the equator. If, for instance, one climbs Kilimanjaro, one observes how at even relatively low altitudes the typical tropical vegetation disappears. The Banana ceases some 2000 metres up and one comes to a wooded zone strongly reminiscent of the leafy forests of the temperate regions. In place of epiphytic flora, long beards of lichen hang from the trees. The climate becomes more and more severe, the leafy woods retreat and we have a zone comparable to the conifer belt of the north. It is represented by the tree-heaths covered with minute flowers, like a veil of flowers thrown over a conifer. Above the tree line one comes upon an alpine vegetation of hardened and contracted flowering plants. Everlasting flowers, especially, abound and give the landscape its character. Finally, vegetation comes to an end in a zone of mosses and lichens corresponding to the tundra. Naked rock follows and the eternal snows.

One of the exciting experiences of a botanist is to ascend a high mountain and to notice how, as he nears the tree limit the vertical forms in the vegetation recede. One after another the plants begin to cling to the ground or the rocks. The trees seem to squat down and as the crowns contract so the trunk and root systems increase. With each higher level the life-forces of the Earth diminish.

The first impression of the ascending mountaineer is the decreasing height of trees and finally the formation of prostrate cushions and mats. The plants shorten their branches, cling closely to the rocks and form dwarf shrubs. Even the herbaceous plants contract their stalks and the leaves draw closer together. Under high alpine conditions cushions and patches of harebells and poppies are to be distinguished from lichens only by their fresh green colour. The same force reveals itself in rosette plants. One need only imagine the following transformation: a plant whose leaves wind in a spiral round its stalk is gradually pressed down; the leaves lie closer together. Finally the leafy spiral becomes like a watch spring, wound in on itself.

Lowland plants transplanted into the mountains show such changes. The shoots become shortened, the leaves contracted and compressed. The clusters of flowers are thereby often drawn close to the ground when normally they would be born on stalks. The root is enlarged, sometimes to fantastic dimensions.

One can study the metamorphosis of plants as a whole as they ascend the mountain: individual forms can be picked out and examined closely; one can study how a lowland form changes when brought into alpine conditions; and even a third approach will lead to similar results. One can study the difference in one species appearing simultaneously in lowland and mountain. Our Valerian serves as an example: it raises its pink umbelliferous flowers to the height of a man. This plant has a relation in the Peruvian Andes which could be compared to a bunch of short-stemmed daisies plucked by a child and left lying on the ground. The juniper, which resembles a cypress in our lowland treescapes, grows no taller than 20 inches on the summit of Mount Etna. Its crown is hemispherical. A forest of them is composed of larger and smaller cupolas. Dense, prickly clusters of needles render it impenetrable – star upon star, a plant reflection of the vault of heaven.

In the alpine landscape, as in the arctic, the flowering forces are

Plate 38: *Silene acaulis* (Moss Campion). This alpine plant with exquisite pink flowers belongs to the Pink family. In contrast to our long-stalked carnations, campions, etc., it forms cushions which consist of densely crowded branches, united in a single massive root stock penetrating deep into the earth.

exceptional. Spring in the mountains appears as if by magic, bringing gentians, primroses, anemones and many other plants, never to be forgotten by anyone who has had the privilege of witnessing it. Hardly has the snow melted when we see the little bells of Soldanella nod in the wind. The contrasts of summer and winter, day and night, frost and heat, light and darkness are sudden in high altitudes. Mountaineer and botanist alike share the same experience; they are faced by the rock masses, and with a heightened sensitivity feel the purity and nearness of the firmament.

The Goetheanist, Henrik Steffens, in his book *Anthropology* compares the vegetation from the pole to the equator to a 'horizontal tree whose roots are directed towards the magnetic poles, and whose leafy crown develops in the tropical equatorial regions ... The same relation which we have followed in this way from pole to tropical region may also be established when we consider the vegetation from mountain altitudes to low lying regions ..., so that we can regard the vegetation from the heights to the lower regions as a reversed tree whose roots turn upwards towards the lichen zone, its trunk being the conifer forests, its branches the broad-leaved woods, and finally its crown the lowest region of the palms.'

We conclude our observations on the metamorphosis of vegetation from pole to equator on the one hand and from the high mountains to the lowlands on the other, by comparing the Earth to two gigantic mountains with their bases set together at the equator. The two poles are like the snow-capped summits. Such pictures will not mean much as long as one sees them against the background of a purely mineral and dead earth. Today we stand at the beginning of a path which is to lead us towards a scientific outlook on the Earth as a living organism. As long as this is not understood, such pictures are meaningless. However, they can become guides towards new laws of a living nature, laws undreamed of considered from the obsolete materialistic viewpoint. The picture speaks the truth no less objectively than the intellect, as we have seen in the plants described. Indeed, living entities are best seen in pictures. If they depict reality and are no mere comparisons, no exact science can raise an objection. On the contrary science will gain from them.

12 Deserts and Steppes

The significance of water for plant life is most strongly felt where it is lacking. Permanent shortage of water creates deserts; the temporary absence of it makes for steppes such as are found in areas with seasonal rains. We often hear deserts spoken of as places entirely devoid of plant growth, but this is not quite correct. We know of a specific desert flora, a flora which is of the greatest interest as it shows nature's means of keeping plants alive under these most unfavourable conditions. The peculiarity of desert plants will be easier to understand if we know the nature of water and what it does in plants.

Wherever it appears, water brings equilibrium to opposing principles. It is a mediator both in general and in the life of individual plants. During rainfall and when the moisture rises again into the atmosphere, water acts as a mediator between cosmos and earth. It gives life to the plant and connects flower with root – the mirror image of cosmos and earth. It creates the leaves and makes them spread wide. We can observe the luxuriant leaf growth developed by plants in wet places and see how water counteracts the form-giving influences of light. The deserts, however, are places with no rain, or very little of it, and that at irregular intervals. Dew is often the only source of life-giving moisture.

We pointed out that the plants of the deserts – like any plant growing in a dry place – have some features in common with the vegetation near the poles or in high altitudes. This relationship is of special interest. If we compare the arctic or the alpine vegetation with plants growing in waterless zones, we soon discover wherein lies the likeness. In both cases the mediating leaf is lacking and the flowers grow out of the contracted and hardened shoot. In the arctic and high mountains the cause is the general configuration of the earth-organism, while in dry areas the lack of water accounts for this phenomenon. In arctic regions water is plentiful but lifeless; on the high crags of mountains it is as good as absent. But in spite of the similarity there is

Plate 39: American Agaves in Sicily.

of course a striking difference of appearance between arctic plants and desert flora.

To describe the general characteristics of desert flora let us consider first a widely known plant: the cactus. Cacti come from the stone deserts of central America; looking almost like stones themselves they are often the only plants a traveller will meet with in these places.

It is quite easy to imagine that the cactus is derived from an ordinary leafy plant, the leaves having merged into the strong, fleshy stem. Cacti are, in fact, fleshy stems. That the leaf is incorporated in them is perhaps not so apparent in the spherical and columnar ones, but shows very clearly in the flat 'leaf-cacti' commonly seen among house plants. Leaf and stem have become one single organ, scalloped or crenellated, with the buds placed in the 'notches'.

The connection becomes still more noticeable when the flowers

Plate 40: Flowering Cactus.

unfold. They do not seem to be part of these coarse fleshy plants but look as if they are stuck on. Looking impartially, one senses a law unfolding before one's eyes. The leaf-bearing shoot, which leads the vegetative parts gradually up to the organs of the flower in a progressive metamorphosis, is lacking in the cactus. Nature makes a leap, and so the cactus becomes the image of a landscape in which the parching heat of the sun lies like a heavy blanket on the dead ground and there is no water to connect the two. The cactus flower shows as clearly as could be wished how powerful the fire element must be. It looks like a flame blazing cholerically, and therein lies the contrast to the arctic flowers which sparkle like crystals. In the cactus the flower is dominated by a metabolism much more intense than that in plants of northern climes. In some cases a single day or night suffices for a flower to burn itself out and wither.

The cactus is the archetype of a whole series of plants which have acquired the same habits. There is for instance in South Africa another 'Cactus flora', but these plants which so closely resemble cacti are Euphorbiae. One sees the difference the moment they break into flower or when, through injury, they exude a milky juice. It is the similarity of climatic conditions which evokes similar shapes in two different groups of plants.

The Aloe, Agave, Echeveria, Gasteria, etcetera, which are favourite house plants, are often called cacti although in their case it is not the stem which grows fat and fleshy but the leaf. They are 'leaf-succulents' in contrast to 'stem-succulents'. Like cacti, these plants store water in their tissue and so create their own small life reserve against the

Plate 41: *Echinocactus minusculus.*

Plate 42: *Euphorbia canariensis*, a thicket of a cactus-like Spurge. The coarse, candelabra-like branches are up to three metres high. Their white milky sap betrays the fact that they are not true cacti. A drop will ooze out even from a pin prick. (Photographed in the Botanical gardens in Palermo).

Plate 43: Cactus Spurge in flower.

hazards of irregular outer conditions. In addition, most of them have a leathery skin and a coating of wax to guard against evaporation. Nearly all succulents are plants of warm and dry climates. But they can also be found in other places, for instance where the soil contains an excess of salt as on the sea shore – (*Salicornia*). Plants find it difficult to take up water from a very salty soil.

The appearance of plants change according to whether a region is completely dry or whether some rain falls now and again. The possibilities are many, and only one has been described so far. Another way for a plant to survive drought is to discard its leaves. A thorny bush in Africa employs this means. The greater part of the year its branches are bare but the moment the rains begin, leaves sprout forth, only to fall off again in the summer drought. So our winter exchanges function with the hot African summer. Representatives of plant families widely distributed in our part of the world (like *Cruciferae*, *Umbelliferae* and *Compositae*, etcetera), which have ventured as far south as the northern Sahara assume a very strange appearance indeed. The stems branch out into tightly-knit round cushions, something like cauliflowers. Long tap roots anchor them deeply and securely in the desert

Plate 44: *Welwitschia mirabilis* on the sandy stony plains of South West Africa. (From Schimper, *Pflanzengeographie auf physiologischer Grundlage*).

ground. A strange example is *Welwitschia mirabilis* of the Kalahari desert. Its barrel-shaped woody stem is deeply imbedded in the ground and only a few centimetres of it project above the surface. Two leathery leaves slashed into many strips are all that the Welwitschia displays. It does not develop beyond this primitive level and could therefore well be considered as a form of giant seedling. The span of the two leaves can be as much as four metres. When the small flowers appear around the top of the stem Welwitschia takes on the appearance of a gigantic *composite* flower lying on the ground. Welwitschia is a true plant of the tropics in that it represents a primitive state of development and produces a flower, while omitting the formation of any true vegetative organs. Its appearance can be understood only if it is regarded as an image of its desert-home.*

* Dr. H. Poppelbaum pointed out that the Welwitschia has a counterpart in the animal kingdom: the ostrich, which for all its size remains a chick throughout its life. Ostrich feathers are down.

Plate 45: A Cactus resembling a pentagonal crystal.

Let us now turn to the steppes. They too can be incredibly transformed in the shortest time. Thoroughly parched areas become green overnight and luxurious plants, often with gorgeous flowers, shoot up from the dried-out soil. In the steppes of Asia, the native land of the tulip, the spring rains conjure up a wealth of flowers which can hardly be imagined; but with the beginning of summer the beauty fades, the bulbs withdraw below ground, the grass dries up and desolation descends again on the land. The following year the miracle repeats itself.

So we see inventive Nature has many possibilities for keeping her children alive in times of drought. What the cactus does with its stem *above* the ground, when it creates pillars, globes, crystal shapes or flat forms, the bulbous plant does in an entirely different way *below* the ground. The bulb is an organ of scaly leaves – or at least segments of leaves. (More about this in the second volume of this work.) When the organs above ground die, the bulb lives on entirely in its underground realm. It is then surrounded by a tough skin and contains moisture as well as nutrients to maintain its further subsistence. Thus it is able to withstand the inclemencies of the climate without damage, as a quasi-individualized plant organ. The cactus too has an individualized organ consisting of the stem and leaves *above* ground. Nor is it dependent on seasonal rhythms like the bulb. It is self-contained within its thick, waxy, leathery skin. Some of the Euphorbiae follow a different pattern. *Euphorbia dendroides*, a plant of the Mediterranean, produces leaves after the first autumn rains and loses them when summer approaches so that the wonderful round bush stands bare, like our trees in winter. Any number of such instances could be given and each would be characteristic of the landscape as well as of the plant. The traveller Farini calls the Karroo in the South African Cape province 'the most terrible, dry, burnt-out, over-roasted, parched, baked, consumed and God-forsaken landscape that the sun ever shone upon'. Such a description could only be given by a man who saw the place during the rainless season, for the first rains perform a miracle. Within a few days green pastures studded with large flowers greet the eye. The desert becomes an unrecognizable wonderland.

Such is the connection between snow and sand and stone deserts. That the barrenness of the deserts cannot always be blamed on the soil conditions, Egyptian history has taught us. If there could be enough rain everywhere on the earth the metamorphosis of the vegetation would continue unbroken from the polar regions to the equator for it is a basic phenomenon – (*Urphenomen*).

13 The Maquis Landscape

The maquis is an evergreen shrubby vegetation of the subtropical Mediterranean. It covers the coast of the Adriatic from Dalmatia to Greece. At its most typical (as for instance on the small island of Lacroma off Trehinje), the shrubs grow several metres high and are overgrown with creepers, making a wilderness of vegetation. Most of the plants have hard leaves like myrtle, arbutus, bay and locust tree or they bear needles or scales like the juniper and the tree heaths. Pine groves, for instance of the delicate Aleppo pine of Dalmatia, and evergreen oaks are part of the landscape.

It is a strange experience to wander over the sun-baked limestone rocks in summer and, nearing the coast, to enter the scrub. Every shade of colour is to be found from light yellowish-green to dark bluish-green. One feels stimulated and refreshed although only the evergreen plants retain their leaves; everything else is dried up. The hardened appearance of all plants seems unusual when compared with shrubs of our temperate climate. They look as if they are fashioned from wire and tin. Many hidden thorns and prickles render the thickets impenetrable and one has to keep to the narrow paths.

Here it becomes evident that light is a formative force which sharpens all parts of the plant and transforms leaves and shoots into thorns. If the light forces are to prevail, water must be withheld, for it would conteract the light forces and produce rounded, bulging, soft shapes. During the rainy season for instance one finds quite different plants even in areas with habitual summer drought. Spring is the time of luxuriant vegetative growth; where there was parched land and bleak vineyards there will be prolific plant growth. Everything becomes green and many lovely flowers, familiar as garden or house plants, cover soil and rocks with colour. The mountains, still snow-capped, look down on a land that is like a garden.

Soon, however, the picture changes again. The rains cease and the temperature, rising daily, reaches furnace heat. The nights bring no

Plate 46: Ancient Olive trees near the temple ruins of Ghirgenty.

relief. Drought sets in. Day after day the sun burns down from an eternally blue sky on the dazzling white limestone rocks. Plantlife subsides into 'summer sleep'. There are no more flowers.

Only the plants which are biologically protected from drying up by their hard leaves or needles keep any foliage. So the colour of the landscape remains green, even though everything around seems to have died off. But this lack of activity is only apparent. Something is happening on a different level. The flooding light and the scorching heat inwardly transform the plants. Instead of outer appearance there is an inner activity. By the withdrawal of water no longer provided by the soil a new situation is created. Now is the time for the olive tree to make oil in its narrow silvery leaves, for the lemon and orange to develop their aroma. The fruit is being prepared. The bay leaf gives off a lovely scent when rubbed between two fingers. The sun-heated leaf is more aromatic than the cold one, for the scent is stronger when the process is still active. By contrast the rock rose (*Cistus crispus*), incomparable gem of the maquis in early summer, is not scented. Everything goes into the making of its paper-thin, white, red or yellow flowers.

When open to the more intimate processes of nature one can experience the cosmic qualities which sink down into the plants. The oils and perfumes are materialized light, substantial heat. The plant is the living organ that collects these heavenly gifts. At noon the activity is most refined. Compared with the strong scent of the European pine woods, this breath of the southern plants is more pure. It is like balsam coming towards you in warm waves; right and left of the path stand rosemary bushes as tall as a man; sage finds its way right up into the mountains, together with other labiates. Each plant exhales its individual aroma.

This transformation process could briefly be described thus: During the spring rains vegetative growth prevails and a mass of herbaceous plants grow and flower. During the following period of drought the process becomes more inward. It could be likened to an invisible second flowering stage which takes place in the green parts of the plants and gives rise to the oily and aromatic substances. Light and heat penetrate deeply into the plant metabolism. During the 'summer rest' vegetative life yields itself to the inflowing power of the cosmos – a thought tinged with reverence and awe. The ultimate receiver of the gifts is man. He harvests the oil and enjoys the delicious fruits and spicy herbs, whilst medicinal plants give him their healing power.

III   The Threefoldness of the Plant and the Picture of Man

14

Before we can discuss further the various stages of development of the plant kingdom we must find some scale or standard by which we can assess the being of the plant. Plants vary greatly in their degree of completeness, the most highly developed ones being the flowering plants. They alone have all parts of the plant organism fully developed and differentiated from each other, namely: root, stem, leaf, flower, seed, fruit. Only the flowering plant has its own organ for each function. In the lower plants, organs and functions are interwoven. One can understand the lower plants only by comparing them with the flowering plant; but one must realize which part of the flowering plant is developed in the lower plant. Therefore it seems natural to take first the complete flowering plant as a sort of 'common pattern' and then look backwards into the various stages of development. This will provide us with the right approach. The plant is a threefold being. the main parts – root, leaf and flower – have already been mentioned. But the plant does not stand in isolation; it is connected in the most diverse ways with the other kingdoms of nature. It is important to see this in the right light and here the threefoldness of the organism as described by Rudolf Steiner is invaluable. The threefoldness of root, leaf and flower corresponds in the human being to head system (nerve-sense system), chest system (breathing-circulatory system) and digestive system.

We see now that we have to consider the plant as man upside down, but not as regards outward appearance of course. Rudolf Steiner speaks of 'functions', not separate parts.

As in the human head, so in the root, hardening, mineralizing tendencies are to be found. The ash content is evidence of this. The root takes its forces from the earth in which it spreads out. Man carries his head erect and rounds it off like a sphere – an external picture of his individualized thought life. He is rooted in the spirit. The plant is rooted in the earth, becomes almost part of it, and takes from it the

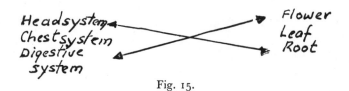

Fig. 15.

necessary minerals. At the same time it is subjugated to the earth's hardening tendencies. We need only picture the plants growing in mountain and polar regions where the earth forces overpower the vegetative ones. Dwarfed, contracted forms result because the root forces penetrate into the vegetative parts.

Through rooting in the earth the plant unites itself with all other plants. The earth being a 'head' for the plants, it unites all plants in it, while man carries his head as an individualized globe. In spite of these differences the tendencies in both cases are the same.

So far we have discussed the plant root only from the point of view of its hardening and mineralizing tendencies. Earlier on we looked at it from quite a different angle. It was considered as the most alive and active pole of the plant. The root could not grow if the earth did not have its own life forces, but in these earth forces there is also the tendency to mineralization and death. The root, once it has passed through its period of living growth, hardens. To understand this we must first consider human development. Man grows from his head, from above downwards; the plant grows from the root upwards. In the embryo the head is developed first; it hurries ahead of the rest of the organism, particularly the limbs, and during the first few years is the centre of the growth forces. During this phase the head has not yet hardened and is only slightly mineralized. The fontanelles remain open as long as it grows. Only after several years, as consciousness develops more, does the head become the centre of the nerve-sense system. So we have in plant and man the same rhythm of development, but the plant is not an independent individual like man in whom the whole organism ages; in the plant root the 'youthfulness' remains in the tips and it renews itself as it grows.

The flower and fruit of a plant correspond to the digestive-metabolic system in the human being (limbs do not exist in the plant). Interesting correspondences are to be seen between plant and man. Henrik Steffens used the picture of the flower sitting like a creature on the green plant which has to nourish it. The flower 'digests' what the vegetative parts prepare for it because it is not able, like the green leaves, to build up organic substances; and this shows the previously mentioned relationship of plant to animal in a new light. The flower can transform substances in the same way as man does his food. The fruit formation of the plant for instance could not be imagined in any other way. Flowers often generate heat, a property which reminds us of blood temperature. The temperature in some cases, as for example in Arum lilies, can be as much as 10°C above the external temperature.

Fig. 16: A SCHEMATIC SUMMARY OF THE THREEFOLD PLANT AS REGARDS FORMATIVE FORCES

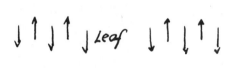

Centrifugal forces (expansion).
Loosening and refining substances, radial or spherical forms.
In scent, optimal rarification and extension of substance. (Warmth process.)

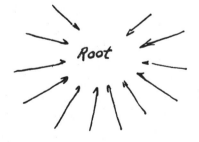

Balance between above and below.
Watery substances and processes interwoven with airy ones. Forms: all transitions from rounded to pointed.
In the uptake of water, in transpiration, in uptake and elimination of gases the 'above' and the 'below' tendencies meet. (Light process.)

Centripetal forces (contraction).
Consolidation of substances to the solid state.
Consolidation of forms.
Suctional, absorptive functions.
(Salt process.)

The combustion, which causes the heat, can be considered as the breathing of the plant – (the word is used metaphorically).

The organs of reproduction are also part of the metabolic system. In man they are centred in the lower part of the organism while in the plant stamens and stigma point upwards. Therefore in this respect, too, man and plant are opposites.

This schematic summary reveals straight away that, suitably modified, it also holds good for man. Here again the basic pattern underlying all earthly organisms is to be seen. According to Rudolf Steiner's description of etheric forces, the centrifugal, expanding forces belong to the warmth ether; the centripetal, substance – producing ones, to the life ether. Between the two stand the light ether working in the airy element, and the chemical ether transforming substances in the watery element.

	Space	*Form*	*State*
Warmth ether	expansion	spherical	heat
Light ether	centrifugal	triangular	gaseous
Chemical ether	suction	halfmoon-shaped	liquid
Life ether	centripetal	rectangular	solid

Rudolf Steiner does not here speak of the hypothetical ether of earlier theories in physics but real formative forces which can be supersensibly perceived.*

When we speak of male and female in the plant it is only a half truth. Reproduction in the plant is not really to be compared with reproduction in man. Human reproduction is arbitrary; plant reproduction is not. It can take place only when the external conditions are suitable. This is shown most clearly in pollination; it is achieved by outside factors like wind or insects. For this reason alone we cannot speak of sexuality in plants. What happens in the plant is completely removed from individual feeling and desires. The plant receives pollen passively; it is only an organ, not an individual or independent person.

It is with the green leafy shoot that we come to the realm where the plant is most truly plant. We have to compare the green leaf with the

* As well as works of Rudolf Steiner see G. Wachsmuth, *Etheric Formative Forces in Cosmos, Earth and Man.* Anthroposophical Publishing Co., London, and Anthroposophic Press, N.Y., 1932.

middle part of man, the chest, where in the lungs the gaseous exchange between the internal and external world takes place. Rudolf Steiner calls this part the rhythmic system.

In the green leaf, fluids meet the external air. It is not without reason that the leaf has been called the lung of the plant. The water rises from the earth via the stem and distributes itself throughout the leaf surface. There it meets the air which enters through the minute pores on the under surface of the leaf. But only in the presence of sunlight can the alchemical processes, the formation of carbohydrates from carbon dioxide and water, take place. To this fundamental plant function some space will have to be devoted.

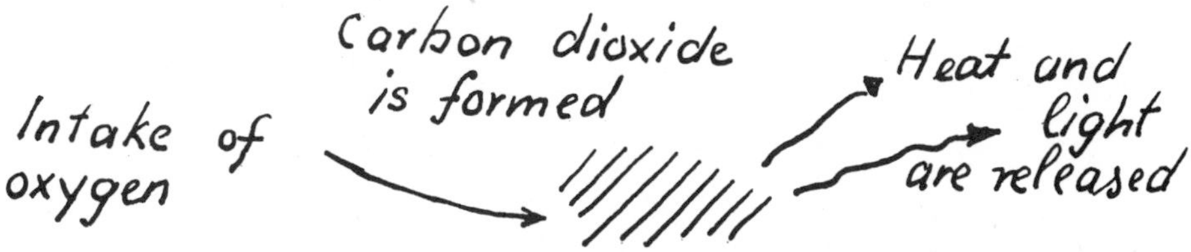

Intake of oxygen

Carbon dioxide is formed

Heat and light are released

Fig. 17.

Carbon dioxide is released where combustion of carbon-containing substances takes place. This may happen in external nature where dead matter is burned, or within the breathing organisms in man or animal. If one disregards water and mineral constituents, practically every organism consists almost entirely of carbon-containing substances. This can easily be proved. If organic substances are heated (not burned) they carbonize (dry distillation). In charcoal production, water and all other volatile substances are removed and carbon alone remains. It undergoes a mineralization process and becomes black. It is different of course if the carbon burns. Then the carbon unites with oxygen and carbon dioxide is formed (CO_2). However, we understand only half if we see only this material process, for in combustion light and heat are set free. These imponderables must have existed in the carbon in a latent form. We do not find them any more in the gas because the oxygen has driven them away and put itself in their place. Light and heat on the one side, oxygen on the other, are incompatible opposites. To the extent that oxygen combines with the substance the imponderables are set free. This is what happens in combustion, (oxidization). Carbon dioxide is entirely void of light. It is cold, a burnt-out husk. As such it is the starting point for the activity of the green plant in the light.

As soon as the air has entered the leaf tissue the carbon dioxide dissolves in the sap. Sunlight works on it, and soon the expulsion of oxygen starts. Plants immersed in water give it off in strings of bubbles.

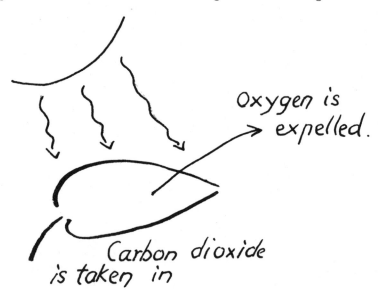

Fig. 18.

The carbon dioxide assimilation is the exact opposite of combustion. In the latter oxygen combines with carbon; light and heat are freed. The plant expels oxygen and fills the empty husk again with light and heat. Carbon dioxide is again reduced to carbon.

However the living plant does not create dead, mineral carbon. It transforms its light-filled archetypal substance and the first material result is starch. Sugars, cellulose (wood) etcetera, also result when water is added as a second factor – (carbohydrates). Other plant substances need the presence of further elements. For instance, protein can be formed only when nitrogen is present. Sulphur and phosphorus are also of decisive importance in most cases.

External combustion is a dead process and only lifeless substance can

116

be subjected to it. The light and heat which are generated are lifeless. The living plant exposed to sunlight reverses this process. It takes in living cosmic light. The result is therefore not only chemical energy but living substance. Chemical analysis which can grasp only dead light and dead matter can never give a satisfactory explanation of these real life activities.

Licht is Liebe . . . Sonnen-Weben	Light is Love . . . Sun-woven
Liebes-Strahlung einer Welt	Radiating love of a world
Schöpferischer Wesenheiten –	of creative beings –
	(Christian Morgenstern)

Breathing in the plant takes place simultaneously with CO_2 assimilation. Seen chemically, it is an internal combustion as we know it in oxygen-breathing creatures like man and animal. The plant does not 'take a breath' but takes up small quantities of oxygen. We have the extraordinary fact that, in the living plant, two polar opposite processes take place at the same time. One, assimilation (photosynthesis), forms oxygen and uses up carbon dioxide; the other, breathing, forms carbon dioxide and uses up oxygen.

Photosynthesis can take place only in light, while breathing is not dependent on light. In the green plant the breathing process is in the background; during the day it is camouflaged by photosynthesis. In non-green plants or parts of plants the situation is different. They 'breathe in' completed organic substances. Parasitic plants get these from their hosts; fungi, yeasts and moulds get it from their nutrient substrata. Roots and flowers get it from the green parts of the plant. The non-green plants have, because of their oxygen breathing, a metabolism more resembling that of man and animal. The typical plant metabolism however is the carbon dioxide assimilation in light (photosynthesis). This exists only in plants. One could therefore call the green plant a 'light-breather'.

Such is the wonderful part played by carbon in nature. Because it is a substance that keeps the balance between the ponderable and the imponderable, between the terrestrial and the cosmic, it can combine with, or separate from, the cosmic with ease. For this reason alone it is more suited than any other substance to be the physical basis of organic beings.

117

It is one of the most interesting properties of the plant that a great deal of what is within man as soul nature has here become form. The plant does not breathe in and out like man, but the structure of the green shoot has a rhythm similar to human breathing. Node follows node, leaf follows leaf and we can consider each repetition as one breath of the plant. We can actually see the rhythm, for the old remains while the new is being formed.

We see, therefore, that the manner of carbon assimilation is different in the two realms of nature. Man absorbs carbon with his food and the blood stream carries it to the entire organism where it assists in building up the body. As the body continually received new food it would eventually grow stiff with the carbon in its structure were it not for the opposing forces of dissolution resulting from the oxygen we breath in. Only the latent heat and light forces in the carbon are retained and transformed into blood-heat, etcetera. Oxygen takes hold of the carbon in man and so lets the body retain its mobility and plasticity.

Comparing man and plant thus, one finds that the plant performs what is not allowed to happen in man. Once the plant has formed a leaf it cannot be 'un-leafed', nor can it scrap a leaf and form another one instead. The plant is 'fixed' within its carbon structure. This would mean death to man.

Observing the growth of the plant one realizes that it can live only by continually forming new organs though the old ones remain. We get to know the 'gesture' of the plant, through which it reveals itself, by looking at it from its first leaf right up to the flower, and by comparing the various forms. They are all there simultaneously. In man one has to observe what happens in time, one thing after another. To do this consciously helps to give an insight into the nature of both plant and man.

Let us return to the rhythmic process in plant and man. Unlike man, the plant does not have its own rhythm. Its sap does not pulsate in a heart rhythm. Its food intake and growth take place within the Earth's rhythm of summer and winter, day and night, rain and sunshine. The rhythmic processes which have become individualized in man have remained external to the plant and are therefore common to all. Plants are only organs; their breath is the wind, their blood the water, the sun their heart.

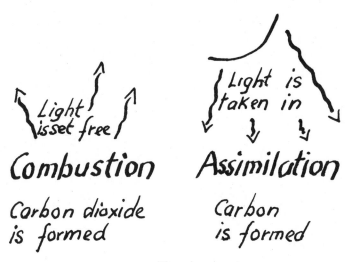

Fig. 19.

Nobody who knows these facts could arrive at the false conclusion that plants have an individual soul. The soul forces in the plant kingdom do not bestow inner feeling, but outer form. These soul forces become a projected picture which through colour and form appeals to our senses. Most striking is the flower, which of all the parts of the plant is closest to our soul. This is why some flowers strike us as symbols of soul qualities. The carnation seems proud, the forget-me-not guileless. Even wishes and desires can be found portrayed, as it were, in the plant kingdom. It can depend on the maturity of man whether to him the rose is beautiful, the lily pure.

The individuality whose soul manifests itself in the plant is the Earth itself. Not the individual plant, but the giant body of the earth, may be compared with a human being. It is not too much to say: if the earth could not send its soul forces into the plant world, the plants could never look as they do and as we love them.

At the poles the earth is 'head' and in those regions it is ruled most strongly by mineralizing and life-destroying forces – just like the human head. The result of these tendencies in the plant has already been discussed. However, the comparison of earth and man can be carried even further. In our head we develop our thought activity. Thinking and the building of mental pictures are possible only because in the head the formative forces have largely withdrawn and are transformed in a higher metamorphosis into soul and spiritual forces. On the earth we have externalized before us what in man is inner soul life; the delicate flowers of the arctic sparkle like crystals, standing as they do on their hardened shoots, a carpet for fairies over the hard earth's crust like the flowering thoughts in our mineralized heads. The arctic flowers are visible thought pictures of the earth in the regions which represent its head.

In the tropics the flowering impulse is – like the soul in the human metabolic system – steeped in the vegetative processes, and so we find the flowers looking like organs, with heavy, earthy, sensual colours and scents. The outer appearance of the plant world itself shows us that the metabolic system of the earth organism is in the tropics, the head system at the poles, and that the temperate zones correspond to the rhythmic system in man.

The physique of some human races shows that man, too, is subjected to a change in appearance or a metamorphosis. As we travel from north to south we encounter first the Eskimo, the inhabitant of the polar regions. He has a large head in comparison with the rest of his body. As we have already mentioned, man is so organized during his childhood that the head grows quicker than the rest of the body; later on this is adjusted as the individuality develops. With this in mind, we can say the Eskimo keeps his childlike physique throughout his life. The races of the equatorial regions, the Negroes, show the opposite phenomenon. Their bodies are larger in relation to their heads; they keep on growing too long, and the limbs shoot out beyond their goal. The result is a limb man, a physically over-developed adult type.

Formative or 'swelling' forces cause the plant metamorphosis. 'Age' or 'childhood' forces form the human body. The most beautiful human form and the most harmoniously formed plants belong to the middle, the temperate region. It is owing to the inherent nature of the earth that the metamorphosis from pole to equator is directed towards the bodily form. Through this, soul and spirit can be influenced.

An entirely different metamorphosis proceeds from east to west. It is directed towards the soul and spirit and thence influences the bodily form. The phenomena of the east-west differences in man lie in the mental realm. The American, as western man, feels strongly connected with the earth which he quite consciously transforms. He is a technician and shows it by his efficiency. The Asiatic, on the contrary,

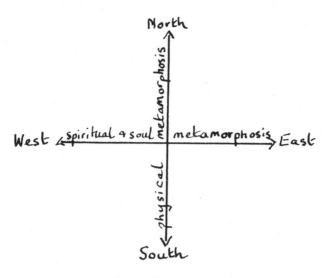

Fig. 20.

considers his incarnation on earth of less importance. His real home is in the supersensible; his soul life is therefore more dream-like when compared with the awakeness of western man.

In the plant realm we cannot speak of soul and spiritual forces as we do in man; but we can study the physiological processes and see if in some hidden way they may show any relation to the mentality of man.

Usteri showed that the influence on man of some cultivated plants can be understood only if seen in connection with the region of the earth where they are at home. For instance, the opium poppy, native of the orient, has the effect of estranging man from the world and driving him into a dream world. The opium smoker loses his will, the youngest and most precious soul activity of man. Dwindling to a skeleton, he becomes a useless member of society. Tobacco is the opposite of the opium poppy and, in accordance with its American home, encourages the earth-directed sense activity. It wakes man up. Thus the soul and spirit in man react to plant processes which otherwise would not reveal themselves.

IV The Ladder of the Plant Kingdom

15

Within the diversity of the plant kingdom, the perfect and the less perfect stand side by side, the higher next to the lower. Here, as in the animal kingdom, we find different grades of organisation all appearing at the same time. The higher a plant's development, the fuller is its rhythm of unfolding and the more distinctly separated are its functions. Each process has its specific organ. Conifers and ferns can teach us some basic facts. A fern obviously stands at a lower level of development as it has not yet a specific organ for the dispersal of pollen, i.e. the flower. This function has been given to the fronds. It is not yet a true link in the plant's development so therefore it cannot be the cause of any profound transformations, nor can it become the ultimate goal of a metamorphosis. The flower alone is the perfect outer image of this function. In it, organ and function are one, while in the fern nature still improvises, as it were; she has not attained her full perfection.

This one instance can rightly direct us. It is impossible to assess the less perfect unless one has acquired the concept of the perfect, i.e. the flowering plant. This is why the flowering plant has been placed before the reader at the beginning of this work. Starting from there, we can see how this archetype becomes simpler step by step in the lower plants and how various organs and processes are combined and integrated. So, if we use the flowering plant as a standard by which to judge all other plants, we shall be following Goethe's method. He looked upon the lower plants as flowering plants which had only partially developed and were therefore imperfect.

It goes without saying that we do not intend to found a new botanical system; we simply want to work out certain basic elements in plant metamorphosis. It may further be asked in what relation our present lower plants stand to the flora of past periods of earth evolution, namely the fossils which have been made so intelligible by the admirable reconstructions of paleontology? We know of ferns, equisetae, clubmosses, algae, etcetera of the present day as well as of the distant past. Both lines of evolution coincide in their main points, but they are not identical. Are our present cryptogams simply the descendants of former plant types, remainders so to speak, or are they novel creations by which nature almost remembers her cosmological past? In the latter case, we should have to assume several independent phases of development of earth evolution, each one of them the outcome of a new act of creation. The basic law of biogenetics, usually applied only to single organisms, would then concern whole stages of development, as shown in this scheme:

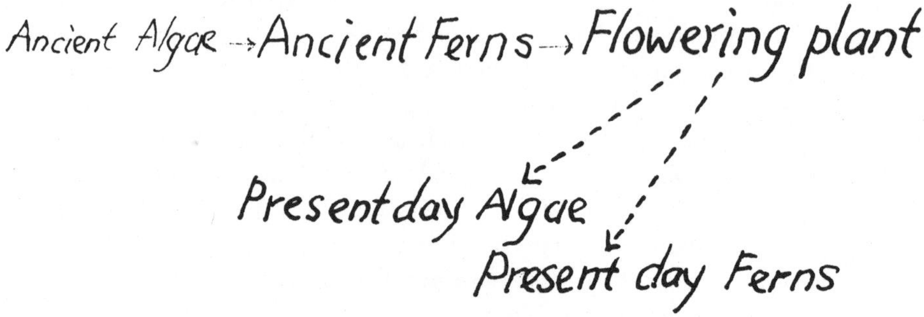

Ancient Algae → Ancient Ferns → Flowering plant

Present day Algae

Present day Ferns

Fig. 21.

The problems connected with this question have weighty implications for our whole conception of nature, but they require more far-reaching views than have been given so far. Outer resemblances are not infallible; in fact, they may even lead into error as we shall see later on. Evolutionary theories will be more widely discussed in the last chapter. Meanwhile we should mention that, because of their resemblance, we have no right to take for granted that our lower plants and the ancient stages of development stand in direct connection with one another according to the theory of direct descent.

16 Conifers

The step from flowering plants (*Angiosperms*) to conifers can present a great many interesting transformations to the student of plant metamorphosis. The difference in the flower is striking. Even among flowering plants there are many either without calix or corolla or with insignificant ones. The conifers, however, show no trace of such organs, and this is why the layman finds it difficult to see that conifers (cone-bearers) are flowering plants. From the botanist's point of view, however, colourful corollas, scent and sweet nectar are not necessary for the making of a flower. He considers the ovary and stamens as the essential parts. When they are developed he speaks of a flower.

Seeds are different. In the conifers, for the first time, seed formation becomes significant. If a ripe cone is taken indoors and left to dry until it opens, hundreds of seeds fall out, looking rather like tiny winged insects.

There is a great difference in seed production between higher plants and conifers. We have already described how the ovule is placed within the ovary in the higher flowering plant. In it, the ripe seed is either enclosed in some kind of capsule or imbedded in a juicy fruit. It is characteristic of these highly developed plants that the ovary is closed (*Angiosperms*). In the conifers the ovules (seeds) are not contained in the ovary. They lie in pairs along the flattened carpel leaf. The seeds of conifers are naked (*Gymnosperms*).

A cone is a small inflorescence not unlike a catkin. It consists of very many carpel leaves which are arranged in beautiful spirals like fish scales. Each one of them represents one flower, much simplified and reduced to one leaf. The seeds become free when the cones disintegrate as in the silver fir, or when they open up like the cone of the spruce.

Let us now try to complete the picture of the conifers from another angle. The difference in mood created by a pine wood and a wood of deciduous trees is striking. The latter seems full of light and sound while the former is still and almost solemn. There eternity seems to

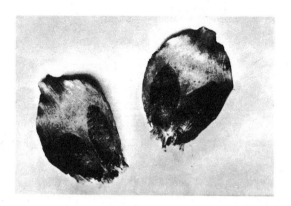

Plate 47: Scales of a Spruce cone, torn off by a squirrel. The impressions left by the two seeds are clearly visible, and also the tip of the scale which in the flowering stage is bent downwards.

Fig. 22.

Plate 48: Seedlings of *Abies alba* (Silver Fir). The cotyledons can still be distinguished between the six-pointed star of the first needles.

speak to us. Woods of broad-leaved trees appear youthful, woods of conifers, ancient.

The difference in shape and formation, too, is considerable. The branches of beech, birch, oak, etcetera spread out into the atmosphere as the veins and nerves of man and animal spread through the body. There is an organic entity which can be felt if we concern ourselves with trees more deeply. How different are the pillar-like trunks of the conifers! They shoot up straight with branches arranged around them in whorls (as in *equisetum*). The older and bigger they get the more their own weight draws them down towards the earth. Gravity takes hold of them. While deciduous trees show organic forms of growth, conifers follow laws of formation found otherwise only in the mineral kingdom. Whorl follows whorl without variation or transition and it is wonderful to see the geometrical figures formed by the branches of young trees. Each whorl is a star composed of crosses.

Although it may seem far-fetched, there is a connection here between two phenomena of nature. For do not these forms remind us of snow crystals, those purest images of the cosmic formative forces in the dead world of the mineral kingdom? The conifers are both

Plate 49: Spruce, Fir and Larch, as seen by an eleven year old child.

terrestrial and cosmic. The formative principles of their growth and also the process of seed formation are cosmic, but the substance in which these forces work are earthly and hardened.

The conifers tend to become rigid. They represent the stem-principle. As they grow older their wooden parts predominate as the Cembra pine of the high mountains shows very clearly. The American Redwood (*Sequoia gigantea*), the tallest existing conifer (up to 140 metres), rises skyward like a huge pillar but its ageing crown is quite negligible. One can say of conifers that even their flowers and fruits are made of wood. The fruit (fir-cone) is the lignified flower.

Yet in spite of all this, and although no colourful blossoms are visible, the conifers are flowering plants. The substances and forces which are normally used up in the formation of blossoms are all there, but are guided into different channels. They appear as resin and as the essential oils in the needles which have an aromatic fragrance. These oils are volatile and combustible like the essential oils of flowers;

resin makes pine wood burn with a bright flame. The flowers of the conifers slumber, as it were, within the tree. If pine wood and needles are set aflame, these awake and reveal themselves. This idea may seem strange, but a flower need not necessarily manifest itself in outer form. It is a process of a certain quality, and is caused by light.

In some cases taste and smell lead us deeper into the secrets of plant life than do mere appearances. The needles of some conifers exude a nectar-like liquid which bees come to collect on warm summer days and to which even butterflies are attracted.

And so our picture of the conifers becomes more and more complete. Hardening and retarding tendencies take hold of the tree and give it its dense appearance. At the same time it is these which make it an evergreen plant by placing it beyond the rhythm of the sun. They take hold of the flowering impulse, hold it fast within the tree and channel it into a different material expression.

It is not surprising, then, that a conifer looks quite distinct from all other plants, even with respect to leaf-metamorphosis. Let us recall all the preparations necessary for the appearance of a flower. The process penetrates the whole green shoot and step by step everything has to be transformed. At last the flower emerges, a real, novel creation even in its material qualities. We have described how the impulse of flower-formation is held back within a conifer tree. This displacement of forces shows its effects on the formation of the tree. Without any change taking place, it puts at the end of its twigs a small model of itself, a model containing carpels and stamens. Each cone is a little tree with the scaly carpels standing around the central spindle in spirals. What is held back in the tree must needs be wanting in the flower. This is why conifers have no beautiful flowers. They go straight into seed production. The process is *added on* to the plant, as it were, not developed out of it. The same applies to the tender little 'pollen-trees' which arrange their stamens around a central stem.

That the 'gesture' of the conifers does not suggest the process of pollination, shows just how primitive the organization is. Nothing of the dispersing tendency so characteristic of the majority of flowers is to be found here. The process cannot be read from the outer appearance of the plant, because its correspondence with the organ has been lost. The conifers show clearly that in them the flower impulse has been abandoned halfway to fulfilment and the apparent hindrances make

Plate 50: Spruce cones in their flowering stage. At this stage the colour of the cones is carmine red and the scales are bent over which makes them look like little caps. Only later, when the cone becomes green and hangs downwards, do they overlap like fish scales. The ripe cones falls off as a whole, as opposed to the Fir where they remain standing upright and shed their scales.

an interesting study. They round off the picture of the conifers, these trees which bear the mark of both the earth and the cosmos.*

What has been explained from the botanical point of view also has a bearing on the significance of the symbolic nature of the Christmas tree. In putting lights on these trees we continue where nature left off. What else can the lights mean but making the tree radiant as if it were

* In his lectures to teachers Rudolf Steiner suggested that the stages of development in the plant kingdom be compared with the different ages of children. The age just before the change of teeth should be compared with the conifers. At this age the soul forces are still largely bound up with the functions of the body, like the flowering impulse in the conifers. They start to become free with the change of teeth and make the child ready to attend school. What appears as the flower in the higher plants can be compared with the emergence of the light of consciousness within the soul of the child, for what happens in the plant can be found in man's evolution at the level of soul and spirit. The birth of the flower has its counter-image in certain qualities of soul and spirit. Both are potentially present before they are apparent, but in a different form and more bound up in the bodily nature. (See *Discussions with Teachers* (11th Discussion), Rudolf Steiner Press, London, 1967.)

Plate 51: Pollen cones of Spruce.

in bloom? The flowering of plants is a light process. In some Alsatian sagas, especially those of Celtic origin, the word 'light' is often used for 'flower'. Flowers are truly born of the light, for trees, enkindled by sunlight, shine in bright colours. 'A tree can only become a flowering flame, man a speaking flame, an animal a moving flame,' says Novalis clairvoyantly in his *Fragments*. And again: 'The flower is the symbol of the mystery of our spirit.' Symbolically speaking, we redeem the fir from its rigid, wood-bound state when we put lights on its hardened twigs. The tree itself is a symbol of mankind become entangled in the meshes of the material world and longingly awaiting redemption. Man's hopes speak from the Christmas tree. Nature denies fruit-bearing to the conifer. Man puts lights on it and having done this he can continue the act of creation and hang red apples, golden nuts, etcetera on its branches. Such symbolism is no mere chance.

17 Ferns, Equisetae and Clubmosses

The ferns can be regarded as a real *stage* in the development of the plant kingdom, even more so than the conifers. Actually they are a definite turning point, and stand in a central position in the plant kingdom. From there we can look both backwards and forwards and observe the transformation. Algae, ferns and flowering plants signify the three great stages in the plant-formation process of the earth which can be seen clearly in our present day flora.

In the *pteridophytes* the natural threefoldness is visible. The ferns are plants which develop the leaf in the most varied and complete form. The horse-tails (*equisetae*), a second group, consist of stems only; leaves are completely absent, but the stems are arranged rhythmically like the leaves in ferns. The third group, the clubmosses (*lycopodiae*), play a very subordinate part in our present flora, but during the carboniferous period they were tremendously developed and formed huge odd-looking trees. According to Alfred Usteri they could be compared with inflorescences without flowers.

So we have at this fern stage all the important parts of the vegetative shoot already developed, stem, leaf and inflorescence. However, though they are all there, they are not yet combined in one whole; each part is a plant all on its own. If one wished to make the fern into a flowering plant one would have to combine the equisetum with the fern and the clubmoss. The fronds of the fern would have to be attached to the stem of the equisetum, one leaf to each node, and the widely branching system of the clubmoss would then have to be added to it, so that this synthesized plant would also have an inflorescence.

It must be emphasized that such a synthesis should not be imagined in a mechanical way. It is only conceivable as a true metamorphosis. An equisetum can be only what it is because no other plant element adorns the stem. Were leaves and inflorescence added, it would lose the very characteristics of its one-sidedness. To be able to imagine such a transformation requires the ability to enter into the realm of living

Plate 52: Common Clubmoss (*Lycopodium clavatum*).

things with one's very thinking. It is an exercise, not an arbitrary game.

That this synthetic combination of equisetum, fern and lycopodium encompasses not only the green shoot of the higher plants but the flower itself, cannot be demonstrated at this point. Further explanations are necessary. With the appearance of a flower the metamorphosis of the leaves begins to take place. In the ferns this is not to be found.

The somewhat daring idea here expounded has been put into practice by nature. The flowering plant *is* this synthesis. The development was not a continuous one. Between definite stages of physical development we have to imagine purely spiritual conditions. All this will be explained in more detail later. Meanwhile we are left with the intriguing task of studying how each single part of the flowering plant makes a whole plant. The analysis of the flowering plant gives us an understanding of ferns, horsetails and clubmosses.

18 Ferns

From the very beginning of its growth the fern can be differentiated from any other plant at a glance. One recognizes it by the way the fronds are curled up – a fact of great significance. It is only necessary to compare the growth of a fern with that of a flowering plant to discover its essential nature. The fern is *all* leaf. There is really no stem. Only the parts which are below the ground are comparable to the axes of the flowering plants. Most of the indigenous ferns form rosettes which grow from underground rhizomes and are often arranged in beautiful funnel shapes. Even bracken, one of the most impressive ones, only *looks* as if it had a stem; in actual fact the shoot is only a leaf. What looks like a stem is the leaf stalk; it has no nodes.

The flowering plant pushes radially out of the ground when it starts to grow. The individual leaf develops its tip first, and the parts nearest the stem come later as the leaf itself grows more tissue. The frond of a fern on the other hand develops in the opposite direction. Instead of growing from below upwards it grows downwards or from the inside outwards, as that part which provides the young frond with tissue lies in the centre of the curled up leaf. It is only a first impression that the frond grows from the soil upwards. On closer observation one perceives that, as it uncurls itself it appears to be pushed down on the earth from above. Herein lies a basic difference from the flowering plant.

All ferns, equisetae and clubmosses have one common characteristic: they bear no flowers. Their highest member is therefore the green leaf which is, however, very elaborately shaped.

Most ferns like shade, and need soil which is damp and rich in humus if they are to grow well. The most beautiful ferns are found in wooded areas where they adorn stones, tree roots and old walls all summer long. They do not grow only for a short period as do so many higher plants which die off after fruiting. The predominating organ, the leaf, is durable and dies off only in winter. Evergreen ferns and 'deciduous' ferns exist only in warm regions.

Plate 53: Bracken.

H. Christ writes in his book *Die Geographie der Farne* (Geography of Ferns) that most ferns are to be considered as *Mesothermic Hygrophytes* (plants which grow in medium temperatures, needing moisture) and only a few as *Xerophytes* (plants which prefer dry soil). The humus of the forest soils plays a major part in fern development. As the ideal condition of fern life on earth he describes the forests of tropical highlands and especially the remarkable moss-forests in Malaya.

Each frond is a many-membered system and only very few, such as the Hart's Tongue, have a simple leaf. If one studies a collection of ferns one is struck by the variety of different leaf formations. The fact can be understood only if one has previously made a study of flowering plants. Then one has an idea of the laws which reveal themselves in the pinnate, pinnatified, lobed, palmate, etcetera formations. If one includes the foreign ferns the possibilities become innumerable. All possibilities occur in a single type, the fern, not merely a few varieties but a whole creation of forms 'where a wealth of divine thoughts is

135

Plate 54: Unfurling frond of Bracken.

manifest, which to our feelings suggests at times even a touch of humour' (H. Christ).

As varied as the leaf formation is the whole type of plant, particularly in warm regions, where the lush growth of the ferns reminds one of carboniferous times. Their life forces are as strong as those of flowering plants and like these they include climbers, creepers, lianas, pitcher plants and many more different types. Even epiphytes are to be found. True trees, however, do not exist among ferns. Their place is taken by the palm-like tree ferns of the tropics and southern hemisphere. Although enormous, their fronds are so delicate that they are one of the most beautiful forms of the plant kingdom.

After getting an impression of the vegetative parts of the fern one can consider the question of reproduction. The naked eye cannot look into the intricate method of reproduction and would have us believe that it is a purely vegetative one in accordance with its leafy nature. Using modern methods we discover most interesting facts. Ferns are flowerless plants, yet they carry a reflection of the flower in their

136

Plate 55: Hart's Tongue, in the last stages of unfurling.

Plate 56: Common Polypody.

reproductive system. When the evolutionary process of the whole earth achieves a higher stage, lower forms are also influenced by it. This fact can be regarded as a general law. It is as if, when the higher stage of the flowering plant has been reached, a new era starts, and the less complete plants are affected by this impulse, and, as will be shown later, the coming stage is foreshadowed before its time, i.e. the reproductive process of the ferns can be regarded as an incomplete

Plate 57: Male Fern. Circle of fronds, arranged like petals in a flower.

Plate 58: Lady Fern.

Plate 59: Tree Ferns in the Dutch East Indies.

image of the reproduction of flowering plants. If we remember that our present day ferns are, so to speak, detached parts of the flowering plant in which creative nature remembers earlier times – the higher has shed the lower – one can understand that a reflection of the flower was projected on to the ferns.

Obviously the organs and processes of the flower are only primitively formed in the ferns. They are lowered to the stage of the green leaf and therefore simplified. This makes the picture of the reproduction of ferns look somewhat complicated, but no less interesting. The frond is quite different on the upper surface from the under surface. It is a true leaf only on the surface which grows towards the light. The surface which looks towards the dark dampness of the ground seems to take something of the vegetative nature into the process of reproduction. In early summer strange spots become visible on the under surface. Later on they turn brown and become more pronounced. These are often important for

Plate 60: Holly Fern.

the identification of the species which in some cases would otherwise be very difficult. In some ferns the entire under surface becomes brown which makes it look charred. If these spots are examined under the microscope they are found to consist of masses of tiny capsules, the *sporangia*, which appear to be part stamens and part seed capsules. When they are ripe, the spores, which look like brown dust, are dispersed. If one places a frond, under-surface down, on a piece of white paper and carefully lifts it after a while, a picture of the leaf is reproduced by the scattered spores. The individual spore is a combination of pollen and seed. In fact the word 'seed-pollen' expresses the nature of this product. The process appears like the dispersal of pollen as we know it in the flower, but its purpose is to reach the damp soil, to germinate like a seed and produce a new plant – though not in a direct way.

Before we follow the development of the spore any further we must look at this 'pollen-seed' formation. It will explain the primitive organization of the fern. The higher plant forms its flower separated from the green vegetative part. Filaments and stamens are clearly separated from pistil and ovary. Pollination occurs in sunlight and may be considered as a result thereof. Before pollen can reach the stigma it

Plate 61 : Oak Fern.

Plate 62: Varying leaves of Wall Rue. One can see the same transformation which in higher plants develops successively.

has to travel through sun-filled air. It then grows on the stigma like the spore in the earth. The flower is a repetition of the soil-grown-plant on a higher level, the ovary a repetition of the root but in the realm of light.

This happens in the ferns in a completely different, in fact opposite, way. These plants do not strive towards the light when they start to

Plate 63: Spleenwort.

Plate 64: Enlarged section of a spore-bearing frond of Male Fern. The spore clusters are still covered with a whitish film like a mushroom cap.

reproduce but down towards darkness and dampness. This is where the spore has to go to fulfil its task; and this marks the fern as an earth-bound creation. On the way from the light sphere to the earth, pollen and seed become one – a spore.

As soon as the spore has reached the earth it germinates. The result is an independent green scale, 1 cm long, often heart shaped, called a prothallium. It, too, shows us that its development is away from the light. It becomes the carrier of a real fertilisation process; on its

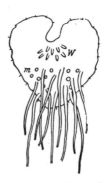

Sprouting spore of Male Fern with the first prothallium cell and the first rootlet. (Much enlarged).

Prothallium of Male Fern. w = archegonia, m = antheridia. (Enlarged).

Fig. 23.

underside antheridia and archegonia are formed. The antheridia discharge antherozoids furnished with tails and these antherozoids move through the watery substrata towards the archegonia and fertilize them. From the fertilized archegonium a new fern grows.

It is a most impressive experience to find fern prothallia in the woods and to realise that here two entirely different plant beings are combined. The prothallium itself is alga-like and undifferentiated; the

Plate 65: Prothallium of Lady Fern with and without plantlet. (4 times natural size.) One can see that the root does not belong to the prothallium but to the Fern, since the upper example which does not show the fern plantlet has no root.

145

Plate 66: *Aphlebia* (From *Grundlinien der Pflanzenmorphologie im Lichte der Palaeontologie* by H. Potonie). Among the ferns of the coal measures there are some which have many small primitive leaflets sprouting from the midrib as well as the normal pinnae (dark in the picture). Thus two different types of growth coexist, one higher and one lower. The significance of this strange fact has not as yet been explained, but does not this resemblance between *Aphlebia* and *Algae* suggest that here we have a stage of plant development where the prothallium climbs up the midrib? Can we recognize the prothallium, though greatly transformed, in the placenta in the ovary? Aphlebia as fossils are widely distributed.

young fern growing from it shows immediately signs of a higher plant with ribbed leaves, etcetera. We are familiar with fertilisation by mobile sperms only in the animal kingdom. There it is in its right place in harmony with the animals' ability to move and feel. We have, therefore, in the prothallium and its development something that would suggest that ferns have some connection with the lower animals. If one follows plant and animal developments back into the far distant past, one finds that they come closer together. The plant incorporates a stage when it can move, the animal fixes itself to the ground (or sea

Plate 67: Hard Fern.

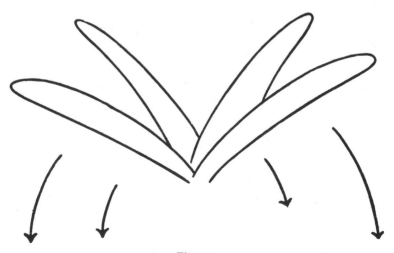

Fig. 24.

Plate 68: Part of barren and spore-bearing frond of Bracken. The difference is obvious as the process of spore formation contracts the edges of the pinnae.

bed). With some very primitive forms of life it is impossible to tell whether they are plant or animal. The animal kingdom in its lower forms 'disappears' as it were into the earth – (worms, unicellular organisms, etcetera) – and so one can understand that the plant, as it nears the earth, can have animal like reproductive processes.

The fern, however, is on the way to getting rid of its animal tendencies. The observable facts of its development can teach us this. It casts out these tendencies, turns them aside, limiting them to a particular stage of development – the prothallium. The fern stands at a decisive turning point in the plant formation process of the earth. The plant frees itself from its lower past and turns towards the light. Only when it reaches the fern stage is the plant about to become itself.* As previously described, it anticipates the entire later plant kingdom, but does not get as far as the flower. Many things suggest that the fern tried to reach the flowering stage, but it would need a new act of creation to achieve it†. Amongst the European ferns there are some which get very close to flowering. The Royal Fern (*Osmunda regalis*), the Moonwort (*Botrychium*), Adder's Tongue (*Ophioglossum vulg.*) separate a part of their leaves and develop them into specialized organs of spore formation. The rest of the leaf remains unaffected. The part which is singled out to bear spores is completely transformed. It contracts and becomes thickly covered with sporangia so that one has the impression of an inflorescence, although it is only a leaf.

Fig. 25:
Adder's
Tongue Fern.

* This is a point from which one can understand why Rudolf Steiner compares ferns with the stage of development of the child when it says 'I' for the first time. In the fern the plant comes to itself. For the first time it forms real leaves and, at least superficially, stems and roots. The frond is nature's image of the 'I' – feeling of the child. (See *Discussions with Teachers* (11th Discussion), Rudolf Steiner Press, London, 1967)

† See remarks in last chapter.

19   Horse-Tails (Equisetae)

Passing on from the ferns to the horse-tails we find that these two are connected in a wonderful way. Each of them actually supplements and completes the other, for each develops what the other lacks. The equisetum is made up of segments of stalks rhythmically spaced and put together. The rhythms continue through the lateral branches right to their very tips. Let us consider the Common Horse-Tail (*Equisetum arvense*) which grows mainly in sandy places, on roadsides, railway embankments etcetera. It somewhat resembles a little Christmas tree. A circle of equal lateral branches radiates from each node giving the plant an almost crystalline appearance. The structure forms a regular star, like a flower, where the various parts too are arranged starwise, even when the leaves on the stem are quite differently ordered. The horse-tail arranges its leaves like petals and repeats this rhythmically the whole way up the shoot. At each node it tries to produce a flower, a corolla.

Nature reveals a consistency in these strange plants which is really astonishing. The unity of plan becomes especially apparent when we compare the spore production in ferns and in horse-tails. At the tip of their green shoots sit the little 'cones' or 'catkins', the organs of spore production. The Common Horse-Tail and Greater Horse-Tail are exceptions. They form separate stems for the dispersal of the spores. These stems appear in spring and are of a brownish colour. Their task done, they die down, and a second set of shoots springs up. These are green and have nothing whatever to do with reproduction. The cone-shaped bodies of the first shoots consist of tiny flat scales which are *regular hexagons* – a visible expression of the formative forces in silica. The horse-tails are silica-encrusted and this gives them their brittle rigidity. When burned to ash the silicious structure remains intact and preserves the characteristics of the living plant in microscopic detail. Silica also exists in inorganic nature in hexagonal bodies; rock-crystal which shows a hexagonal cross-section is pure silicic acid.

Plate 69: Field Horsetails. Right: first shoots with spore-bearing cones. Left: green barren shoots.

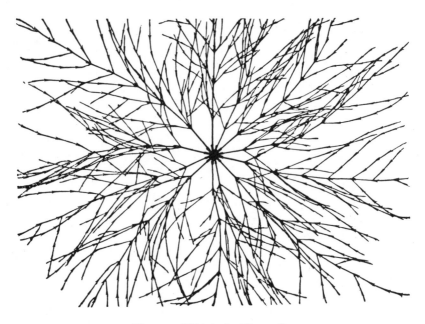

Plate 70: Whorl of a Horsetail.

On the inner side of the little hexagonal scales of the horse-tail we find six little white capsules, each containing numerous spores. In dry weather these spores drift off on the breeze like a thin vapour.

The organs of spore-production in ferns and horse-tails make the polarity of the two very apparent. The fern which consists of nothing but green leaves sets out to represent the leafy component of a flower. The fronds of many species are even arranged in a circle, as if in imitation of a corolla. In the process of dissemination of spores (which has been taken over by the fronds) we find something like a transformation of a petal into a stamen. In bracken, for instance, the leaf margins are slightly curled and in the little hollow space so provided the spores are formed. Here the affinity with the formation of a stamen is unmistakable.

The Hard Fern goes a step further. It differentiates barren fronds from fertile ones. The barren fronds spread themselves in a rosette and are often quite prostrate, while the fertile ones stand erect in the centre of the clump. They are taller and more slender and their pinnea get contracted in the process of spore-formation. The likeness to a flower is

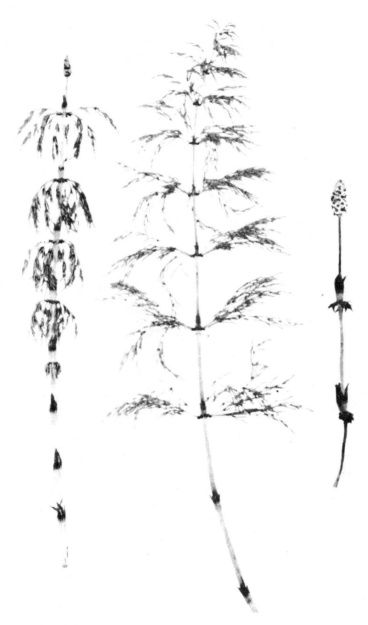

Plate 71: Wood Horsetail. Centre: a barren shoot. Right: young spore-bearing shoot. Left: the same, fully developed. Here barren and fruit-bearing shoots are combined. The spore-bearing shoot subsequently develops green foliage.

Plate 72: Spring shoot (spore cone) of the Great Horse-tail, *Equisetum maximum*. (Natural size.)

even greater. The purely vegetative fronds correspond to the corolla, the fertile ones to the stamens. The fern uses the leaf element to imitate the flower.

Not so the horse-tail. It creates instead an image of that part of the flower which is the direct continuation of the stem. The hexagonal scales with their capsules at the end of the stalk – a lengthened flower spindle – are arranged like seeds in a cone.

The picture is now complete. We see fern and horse-tail each develop *one* of the two essential elements of plant life: leaf and stem. Each makes a complete separate plant. Nor is that all, for even the flower is divided into its two principles. The fern devises its spore-formation on the principle found in stamens whilst the stalky horse-tail arranges its spore capsules in imitation of the ovary.* Could fern

* See Chapter 6, Leaf and Stem Tendencies as seen in the Flower.

and horse-tail be rolled into one the result would be not only the green shoot but even the flower itself of the higher plant.

We cannot gaze without wonder into these mysteries of nature. The invincible law triumphs and can even be found where superficial observation fails to see it. Here it is revealed in the division of leaf and stem tendencies into two separate plants and the transposition of the flower into the realm of the leaf.

Yet the equisetum would not be a pteridophyte if it did not have the same rhythm of development as the fern. They each form a prothallium between each generation, but the equisetum goes a step further. The prothallium of the fern is hermaphrodite; spermatozoids and egg cells are produced on the same plant and self-fertilization can take place. The equisetum has both 'male' and 'female' prothallia though they are very similar in form. They are not leaf-like as in the fern but somewhat antler-like, and show the stalk tendency of the horse-tail. Those which carry the archegonia are somewhat larger than those which carry the antheridia.

The next step would be the differentiation of male and female spores. The present day equisetum does not go this far, but the much more highly developed fossil equisetaceae actually achieved this differentiation (see Chapter 26). This further evolution, however, leads up to the development of seed-producing plants. The female prothallium becomes fixed and stays with its plant, whereas the male is gradually transformed into pollen. Consequently the plant has two principles and pollination becomes necessary. At the same time the processes of propagation are being lifted up into the realm of light and away from the watery earth. The creatures of warmth and light, the insects, become involved. This new metamorphosis which is of deep significance will occupy us later on.

20   Algae

In flowering plants, in ferns, and now a third time in algae, we find a class of plants in which the leaf can be shaped and transformed in every conceivable way. It is especially true of the seaweeds that they are but leaves, never-flowering leaves, floating in the water without roots. These leaves are of a lower order than ferns. It is in fact more correct to speak only of leaf *shapes* for the inner organization remains far behind that of the higher plants.

If one looks at seaweeds with an open mind, one might think oneself in a dream world, they are so unreal. When taken from the water or washed ashore they collapse into shapeless masses as they have no supporting frame. Only the water supports them and only in the water are they able to spread out and unfold. Here there is a world of magic. In the clear water we see a second, underwater plant-world swaying rhythmically in the waves. In places it is like meadows and fields in the upper world, elsewhere it resembles bush or forest. Little moss-like forms mingle with enormous palm-like growths and flabby indeterminate plants. Some seaweeds are brown or even red, but all of them assimilate carbon dioxide in true plant fashion.

Seaweeds are fastened to the bottom of the sea with special organs and so are bound to a definite locality, but often they are torn loose in heavy seas and come to the surface where they accumulate and drift as a kind of seaweed island, sometimes of enormous dimensions. The floating gulf-weed of the Sargasso Sea became famous and caught the imagination of sailors of old. Ships could be in danger of getting stuck in it.

The seaweed represents a stage when only the middle part of a plant is developed. It is cut off above and below; what fasten it to the rocks are not roots and it has no flowers either. Jelly-fish and such like drift in and out among them like dull, torpid butterflies around flowers.

It would not be correct to say that algae have stalks. A stalk is associated with a root and radial growth. What look like stalks are

Plate 73: Herbarium specimens of brown seaweeds from Heligoland (*Punctaria*). (Natural size).

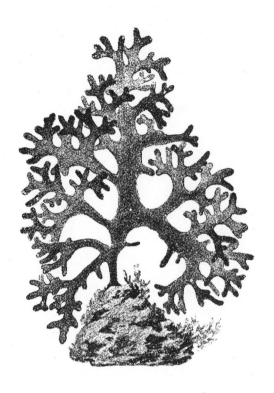

Plate 74: The red seaweed, (*Nemastoma cervicornis*). (From Berthold).

extended or stringy parts of leaves. One has only to picture what changes the plant-world of dry land would have to undergo if it were submerged in water. There could be no more flowers and no more tendency to stiff, sharply defined forms; everything would become yielding and limp. With this would disappear the finer differentiations, clear outlines and leaf edges, etcetera, giving place to indeterminate contours and ill-defined shapes. As the higher plants can be understood only as the result of the interplay of light and gravity, so the world of the algae is determined by the formative forces of water. One might call the algae 'plant phantoms' born of water. They are not yet real plants.

The concepts we have gained from the study of ferns are very helpful for understanding the nature of seaweeds. In fact, without them, it would hardly be possible to present the algae in their true light. In the ferns two completely different stages of development meet. The fern itself is a higher plant than the prothallus which represents in every respect a lower stage of development. The prothallus is only a bearer of the regular fertilization process and must achieve its function in close proximity to the soil.

Algae are in fact nothing other than highly developed prothallia, displayed in an enormous variety of forms. When the fern develops its prothallus it lays aside its own lower past. It quickly recapitulates this stage of development in a rudimentary form and then goes on in the opposite direction to appear on a higher level.

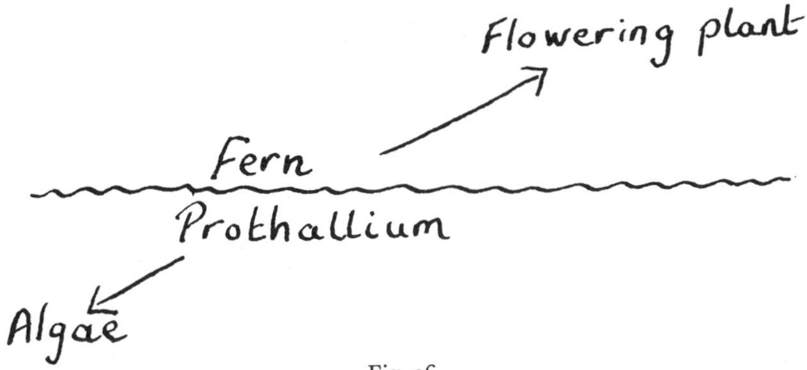

Fig. 26.

If we grasp this turning-point we understand why prothallium and fern are separate and are not simply a unity. Here is a borderline between two worlds and there is no transition stage. At this point one stage allows us a glimpse into the past of the Earth, the other points to the future, even anticipating somewhat beyond the present time. Between algae and ferns there is no transition. For the first time we see clearly the great law: that nature does not develop in a continuous line, but in leaps. The prothallium is a descendant of the algae. The fern is a major new development.

The correspondence of algae and prothallia is quite clear. In the propagation of both we find archegoniae (ovules) and antherozoids (mobile spermatozoids) as well as purely vegetative propagation. However involved – and even partly unclarified – propagation is in algae, it certainly may be compared with the so-called sexual one of the ferns.

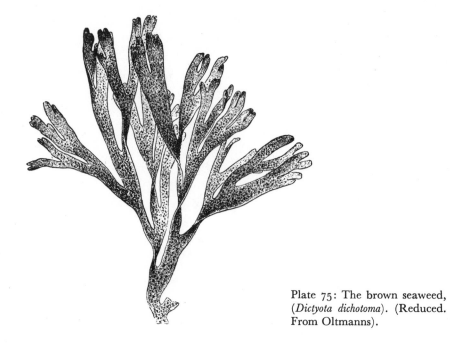

Plate 75: The brown seaweed, (*Dictyota dichotoma*). (Reduced. From Oltmanns).

There is also the characteristic dichotomous branching of algae – spreading out by simply splitting one branch into two of equal prominence. This is the oldest and so to speak archetypal method of plant propagation. It is rare in ferns and only the prothallium resorts to it. The very first little frond which sprouts from the prothallium is often twofold. All the following fronds 'uncoil'. The fern briefly recapitulates the alga state. Division by forking occurs again in the germination of dicotyledons. A significant example of this is the mistletoe. Whenever this branching occurs, it is reminiscent of very ancient plant life.

As already stated in another context, we should go wildly wrong if we imagined that the algae of today are similar to what they were when the whole plant world of the Earth was at the algae stage – when algae were the highest development of plant life. They changed with the changing conditions and we are left with very little to go by if we want to imagine vegetable life in ancient times.

The algae show how lower plants can become mirrors of the impulses in higher ones. Stems and leaves of the higher plants are copied on a lower level, even the veins of the higher leaf can be copied,

Plate 76: The red seaweed, (*Delesseria sanguinea*), with shoots resembling leaves of flowering plants. (Half natural size. From Strassburger).

and some algae have fruit-like additions – bladders. But all this is merely a resemblance in shape.

Such thoughts are bound to occur to the unbiased mind. The palaeo-botanist H. Potanie writes in his book *Grundlinien der Pflanzenmorphologie im Lichte der Palaeontologie*: 'It should be pointed out once more that the seaweed of today is by no means the ancestor of our ferns but at most the ancestor of those algae, especially the forked ones, which may have been more or less like our present algae. That is if the seaweed has not changed its appearance since paleozoic times.' And later on: 'That which palaeontology could say about the phylogenetic origin of the ferns unfortunately has to be rather hypothetical, since the fossils as well as more recent forms of plant life show no transitional stages which could provide a satisfactory bridge from ferns to lower forms of life.'

It is just the lack of these transitional stages that is important. We shall speak in the last chapter of the meaning of these leaps in evolution.

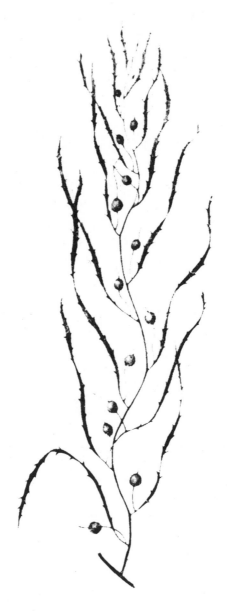

Plate 77: The Bladderwrack (*Sargassum bacciferum*) with berry-like floating bladders. (From Kützing).

21 Fungi

One is apt to think of fungi and mushrooms as either poisonous and to
be feared, or as edible and to be enjoyed for their flavour, like the
capped mushrooms, morels, puffballs, etcetera, fruit of the forest floor
and the meadows. They spring up from a subterranean web of proliferating threads, the mycelium, which is the real fungus, for there are
many fungi which do not make their presence known through a
fruiting body. Mould, for instance, appears only as the well known
greyish-green velvet on jam etcetera. Yeast is still more primitive; the
cells of the mycelium do not even hang together in threads but fall
apart after budding. Yeast has no special organs for propagation;
mycelium and fruiting body are one. Both together are but one cell.

Fig. 27: Proliferating yeast cells – the
simplest form of plant growth at the
fungus level.

The world of fungi is a very strange one. Even the outer form of
fungi and mushrooms is strange when compared with that of ordinary
green plants, let alone their inner construction. Fungi consist only of
threads which keep on growing at their tips and which can get
interwoven and matted together and so form spongy and even quite
firm fruiting bodies. Consequently we cannot speak of 'tissue' in the
sense in which we use the word when referring to those plants in which
cells are closely joined together on all sides. The yeasts are like points
or spots, the moulds are thread-like and only the fruiting bodies of the
higher fungi are three dimensional. The second dimension, the plane,

163

does not occur in the world of fungi. Only the green, leafy plant displays it. The fungi are a world apart.

In order to exist fungi need organic substance synthesized by animals or by other plants, because the non-green organisms lack the ability to assimilate carbon-dioxide. They can therefore exist and be cultivated independently of sunlight in total darkness as the edible mushroom is. The metabolism of fungi can be likened to that of animals, as it is destructive. The end is the total mineralisation of the organic substances. In this the various kinds are able to work together in teams, one variety continuing the destructive process where the other leaves off. Only when understood as a unit can these complicated chain systems show their true functions which, taken all together, constitute the metabolism of the fungi.

The green, assimilating plant raises dead, mineral substances into the realm of life, building up first carbohydrates, then proteins, fats, etcetera. The digestive activity of the fungi runs in exactly the opposite direction; it leads substances back – step by step – into the mineral state.

If weather conditions are favourable, fungi can spring up from the soil overnight; but as fast as they come they can disappear and dissolve again. They are nothing but a metabolic process. No firm frame supports them, no hard shell protects them, there is no kernel or pip within them. It is, however, necessary to try to compare fungi with the complete flowering plant. It is easy to see that they must be related to the metabolic function of the plant which is in the flower and fruit.

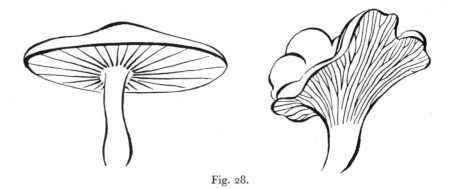

Fig. 28.

The flower has to be nourished by the green parts, as it cannot take up carbon dioxide. So too must the fruit. The words 'fruiting bodies' are very appropriate, for they serve the propagation of the fungi.

The fertile part of the fruiting body is the underside of the cap. It can be laminated as in Agarics, or can look as though pricked by a hundred needles as in the Boletus. The dust-like spores occur either between the laminae or in the pores. As with ferns, the pattern of the underside appears if it is left for a while lying on white paper so that the spores fall out. Some fungi, as for instance the puffballs, are globular or nearly globular and form their spores inside. When you tread on them, they burst and discharge a dusty coloured cloud of spores. Particularly fascinating are those fungi which are not only

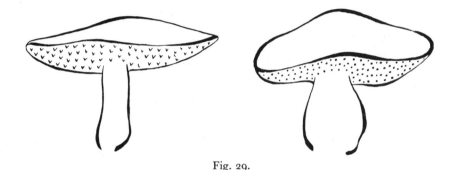

Fig. 29.

colourful like flowers, but even funnel-shaped like a corolla. Their outer appearance points towards their being of the nature of fruit and flower combined.

From its darkness the Earth flowers and fruits through its fungi. The processes which normally terminate vegetative plant growth in the light, i.e. flowering and fruiting, have here sunk down into the soil and become independent. The process takes place at a different level. These fungi sometimes look like shapes born of a nightmare or coming from an underworld. Some seem harmless enough, others have frighteningly glaring colours. There is any amount of variety. Enticement and destruction seem to be embodied in them side by side, and they exert a strange fascination.

It is the task of the spore to return to the forest floor and produce a new mycelium. Nothing like a prothallus comes about in the process.

Fungi remain even below the stage of development characterized by the prothallus-fern relationship.

Were the cosmic formative forces – represented by sunlight – to take hold of the fungus, it would be lifted up from the soil by a stalk and then develop leaves like any green plant. At the same time it would also have to turn its cap upside down so as to make the reproductive part directed upwards to the light. This would make it correspond to the picture of the flowering plant.

For the fungus, however, the soil itself is its habitat. It is born of the Earth. Rudolf Steiner drew the following mental picture for the teachers of the Waldorf School.* He said one could compare the Earth

Fig. 30.

with an enormous tree. 'An ordinary tree grows leaves and fruit. A tree-*trunk* is just soil, pushed up like a heap. For the fungi, soil itself is the trunk on which they grow, for they do not push up an extra trunk for themselves. The tree on which fungi grow has been kept down in the earth. Only the fruits stick out of it.' These words complete the picture we have built up.

Many things point towards the fact that in fungi only earthly influences work, but not the cosmic light forces. Instead of the ethereal scent of a flower we can smell anything from spicy meat to a repulsive stench. Instead of bees and butterflies, we see flies, beetles and slugs feeding on the slimy substances of fungi and the decomposing by-

* *Discussions with Teachers* (10th Discussion), Rudolf Steiner Press, London, 1967.

Fig. 31.

products. What in the green plant is the result of the penetration by the extraterrestrial formative influences of light is here drawn down into the sphere of purely earthly life processes.*

* In human development there is a time when the nerve-senses system which corresponds with the plant root is in a condition in which it is not yet the mediator for conscious perception and mental images. It is then still the centre of processes of growth, it grows itself and is therefore not free of a certain metabolism. This condition – the earliest babyhood – could be called the fungus condition of man. Compared with adult man a newborn baby is 'all head' – a head full of the power of growth.

In the lectures for teachers Rudolf Steiner spoke about the first botany lessons given to children (nine to ten years old) and gave vivid descriptions of how to convey to the child a feeling of what fungi are by comparing them to baby sisters and brothers who can do nothing but drink and sleep. The fungi can do nothing but grow and digest. They can make neither leaves nor stalks nor roots 'because the sun does not bother with them'.

In such ways nature does not become soulless for man, and yet one remains objective, without falling into sentimentality. (Compare footnotes on pp. 130 and 149.)

22 Parasitic Fungi

On no other level does the plant kingdom show so great a variety of parasites as among the fungi. This is not surprising; even the non-parastic fungi need a soil which is rich in organic matter. On dying trees we find fungi which grow to enormous size and gradually we come down to those varieties whose mycelium penetrates the living tissue of leaves, stalks, flowers and fruit, thereby exhausting and destroying them or at least diverting their course of development. Rust in cereals and grasses, mints, mallows, coffee plants, etcetera, mildew and the various fungus diseases of fruit trees, potatoes and vines are rather like moulds, with one difference: they penetrate living substance instead of dead matter.

The cause of fungus diseases is fundamentally a question of level. Until we realize what makes a plant become soil for another plant there is not much point in searching for possible causes. We find that fungus diseases have very little to do with an individual plant or the chance appearance of an infecting spore but rather with the condition of the soil in which the affected plant grows. The parasite must find a suitable nutritive medium. A wet winter and wet spring, for instance, can go a long way towards creating the right conditions. The treatment of the soil, too, its structure and composition can be responsible for fungus diseases. It is a well known fact that there is a natural resistance in healthy plants which are not overbred. But be that as it may, here is not the place to consider the problem from the point of view of agriculture or horticulture, important though it is for those who work on the land. We shall describe the parasitic fungi only so far as they form part of the picture of the plant world.

Seen in connection with the plant they live on, there is no doubt that the two belong together, not unlike flowers and insects. Each plant variety has its own special parasites, but certain conditions are necessary for their development. After a very wet winter the soil possesses an excess of growth forces. As a result, this vitality of the soil – usually confined to the region of the roots – rises up into the green part of the plant. An invisible second 'soil-level' is formed, as it were. The green plant cannot cope with these exuberant growth forces, and a second, i.e. a parasitic, plant growth occurs.

This picture can be further elaborated. One of the characteristics of fungi is their tendency to disintegrate into cells. Even the spore formation must be understood as such a process of disintegration. If mould forms on the surface of a nutritive medium – a piece of stale bread for instance – all it really does is to disintegrate into millions of cells. It turns to dust. The mycelium appears, forms branches and spore after spore drops off their ends. The same is done by parasitic fungi. They invade the host plant, blighting it with their disintegrating tendency. Rust on leaves and stalks, for instance, can well be likened to

Fig. 32: Conidiophore of *Aspercillus* (mould). The spores detach themselves from the top and the fungus disintegrates into single cells. (Much enlarged.)

patches of mould. In the pollen-producing organ – the flower – the parasite simply takes the place of the pollen itself, thereby rendering the flower infertile. The stamens no longer produce pollen but spores are produced instead. This instance is of special interest as it goes to show the functions of the host plant and the tendencies of the parastic fungus can coincide, and the fungus simply takes the place of the normal process without even causing any profound change in the plant's appearance.

If, however, a fungus attacks the green parts of a plant, there a process takes place which rightly belongs only to the flower. A change of level has taken place. From the point of view of the plant, the process has been shifted to a lower level. As seen from the soil level, it has risen. The fungus which has risen from the soil level so enhances the pollen process that in a sense it appears prematurely in the plant. It is a kind of fructification, an abnormality which hinders the normal

Plate 78: Ergot of Rye. The fungus takes the place of the grain. It develops partly in the soil and partly in the plant.

process in the flower, because it anticipates and forestalls it. This process is normal in ferns and has to do with the place they occupy in the course of evolution. The spore-formation is like rust but in the right place. The outer appearance of both processes is strikingly alike. In both cases the underside only of the leaf is affected and the similarity continues into almost microscopic detail. It is of no importance that the spores are formed by the plant itself in one case and not in the other. It is the process which concerns us and not the means which are employed to bring it about.

It would not be wrong to say of a plant affected with rust that it has reverted to the fern stage; indeed this way of looking at it is in keeping with the idea of the displacement of levels. The fern is still deeply connected with the vitality of the Earth itself, and its process of fructification (related to that of the fungus), therefore takes place in the leaves. It has no flowers. If a higher plant gets blighted by a parasitic fungus, the level of soil vitality rises up and the plant is forced down to the level of the ferns, where the fructification process takes place in the leaf.

These considerations help us to a deeper understanding of the horsetail, which is the natural counterpart of the fern in that it develops the stem only. It is all stalk, and this is what the fungus would need to have were it to be transformed into a green flowering plant. The horsetail lifts up the process of spore formation from the level of the soil to the very tips of its stem. In some cases (Field Horsetail, Giant Horsetail), it even transfers it to a special early shoot reminiscent of a fungus. Therefore it is understandable that in Bio-Dynamic agriculture the horsetail is used in the treatment of fungus diseases.

It would lead us too far were we to describe in detail individual species of parasites. Their complexity and enormous variety, however, make them one of the most important objects of study. It becomes increasingly clear that the plant kingdom cannot be understood unless the life of the whole Earth is taken into account.

Let us return to the idea in the previous chapter of the metamorphosis of the mushroom into a flowering plant. The plant proper, i.e. the leaf-bearing stem, does not develop in the mushroom. It sits immediately on the surface of the ground and all its functions are directed downwards. Were it to become a flower it would have to turn itself upside down and look away from the earth. The pollination would then be directed upwards instead of earthwards. All processes would rise up from moisture and darkness to a life in light and air like those of the flowers of the higher plant.

Were this achieved, further outer-terrestrial influences could start working on the newly created plant. An upright stem would overcome gravity and lead to the existence of the very organ Goethe called the 'spiritual staff'. The green leaves would grow out of the stem winding their way up along it according to their various laws: some in pairs (stinging nettles), some in spirals (rose, etcetera). The variety is enormous and the laws are not of the Earth. They are an expression of the movements whose patterns are to be found in the starry heavens. Metaphorically speaking, the plants copy what they perceive goes on in the sky.

With the rays of sunlight the plant receives formative forces which it brings to expression in its outer form. We must look upon sunlight not as a mere source of chemical energy – though materialistic thought tries to do so. The outer light is something eminently creative. It is the medium for the whole wealth of creative form-giving forces streaming down to earth from the cosmos. To see what the light must mean for the plant form, we need think only of the rank shoots of potatoes kept in the dark. *The heavenly bodies are the great teachers of the plant world.* They lay down the laws whereby the naturally formless material is endowed with individual characteristic forms, down to the last detail.

These form-giving forces also have a polar opposite – a principle which sets itself against all laws of formation. Although it has a life of its own, it is not something which makes the plant an image of cosmic events. This self-willed principle is the cell. In no circumstances should we regard an organism as the sum total of its cells. On the contrary – and this should be emphasized – the higher plant exists *in spite of its*

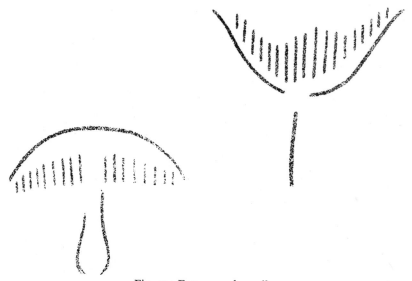

Fig. 33: Fungus and corolla.

cells. Could cells left to themselves be expected to end up as an organism? Has a house ever been known to go up by bricks being piled in a heap? The architect has to draw up a plan according to which the bricks are laid in certain places, numbers and kinds. Some have to be dressed and prepared to fill definite places, or the result would be no proper house.

Admittedly, the plant is built of *living* 'bricks', yet not without a plan – a cosmic plan. The sheer development of cells does not make an organism. Rudolf Steiner once pointed out that the cells *interfere* with the higher form-giving principles. If the relationship is clearly seen

172

Plate 79: Herbarium specimens of *Euphorbia cyparissias* (a spurge). Centre: a normally developed, healthy plant side by side with a deformed shoot. The deformation is caused by a fungus, *Uromyces pisi*. Extreme right: another fungus-infested plant, extreme left: faciated stems.

between life-bearing, malleable material and formative forces it will be found that the picture of the heavenly bodies as teachers of plant growth expresses a reality.

The fungi are plants in which cell growth predominates. The formative forces are lacking and the cell principle develops so strongly that it even destroys the form.

The changes brought about in plants by parasitic fungi show the same thing. The process of pollen dispersal becomes falsified and the pattern which was to be realised in a systematic evolution is in many cases completely annulled. The formative forces can be deflected.

All influences which strengthen the earthly forces as opposed to the cosmic ones, help to bring about fungus diseases. It can be extraordinarily interesting and worthwhile to study the various types of deformity plants are subject to when they fall prey to fungus diseases. A few instances must suffice. A well known plant, Shepherd's Purse, very

173

easily becomes host to parasites. The flower-stalk, penetrated by the mycelium of the alien organism, becomes fleshy and thick, loses its healthy green colour and becomes barren. Members of the Euphorbia family suffer acute deformities. A normal Euphorbia (spurge) is a herbaceous plant with alternate leaves. At a certain point a concentration of leaves takes place. They form a whorl, out of the centre of

Plate 80: Witch's Broom on a Silver Fir – a miniature tree growing on a tree. This is caused by a rust fungus (*Melampsorella*). The needles of the Witch's Broom are pale green. On the underside they are yellow, owing to the spore clusters of the fungus.

which arise several equal branches. Each of them is crowned by flowers as in an umbel and the many inflorescences give the impression of a true umbel. Consecutive form principles take over step by step in order to create such a plant type. When a fungus attacks the spurge, the planned development does not take place. The pattern is blotted out and we see yellowish-green stalks with stunted remains of leaves. Flowers do not develop. It would hardly be possible to identify such a plant as a spurge if one had no previous experience of it.

Finally let us take the Witches' Brooms on trees. They too, are caused by fungi. Each tree has an inherent form principle; trunk and branches grow in a way characteristic of each species and bare trees can be identified by the formation of their trunks and crowns. When a fungus attacks, there is chaos. Branches grow in wild confusion and no plan can be recognized any longer.

23   Bacteria

This part of the fungus flora has become known to man only comparatively recently. Bacteria are almost ultra microscopic and cannot be described without aid of optical instruments. With their discovery a new chapter of natural history began. The pathogenic bacteria became a focal point of man's interest. Many exaggerated ideas and errors followed. Today, however, many voices are raised cautioning us against the belief that bacteria are the primary cause of disease. A few examples can lead us towards a new way of looking at bacteria.

Bacteria are unicellular plants. Some are small rods (*bacilli*), some are spherical (*cocci*), others tapering (*vibrona*), still others corkscrew-shaped (*spirochetes* and *spirilla*). Mobile varieties have minute *flagella*. It is unnecessary to describe them all in detail as everyone knows about these micro-organisms and has seen them illustrated.

Bacteria propagate by simple division. A cell becomes constricted round the middle and eventually falls asunder. In this manner whole chains or other groupings are formed. If conditions are unfavourable (if the nutritive medium dries up or becomes too acid), many varieties start producing spores. In this case they surround themselves with durable shells and can then withstand extremes of heat and cold. Some can stand hours of boiling. The moment the spore finds favourable conditions, its skin swells and germination can begin. Division sets in very quickly and within a few hours many thousands of cells have come into existence.

These facts caused great astonishment when they first became known. Grotesque comparisons with human population expansion were indulged in. It was easy to cause astonishment with astronomical numbers. However, we have no right to compare one cell of a bacteria with a whole human being. Can one speak of birth and death or reproduction, when Nature annihilates hundreds of thousands in one sweep? Such ideas could come into people's minds only because the concept of the whole Earth as an organism is missing in our scientific view of the world.

A bacteria cell is nothing but a cell, and can be compared only with a cell of an organism such as man. What is erroneously described as propagation is actually the same as growth by cell division in any plant, animal or human tissue. Propagation in bacteria is identical with growth of tissue. The only difference is that bacteria let their 'tissue' disintegrate into its parts, i.e. the single cell. As man possesses a stomach, liver, etcetera, as organs for his bodily functions, so, too, the Earth has distinct organs, with the difference that their functions are carried out by single cells – bacteria. They are responsible for the wide range of processes like fermentation, putrefaction and decomposition which constitute the digestion of the Earth.

A few examples must suffice. When milk becomes sour through the action of various types of lactic acid bacilli, or when quantities of dead wood on the forest floor are broken down into their mineral components through bacteria, these are processes by which substances are given back to the mineral part of the Earth. The reversion of the organic to the inorganic state is the main task of bacteria, and a great variety of them have to work together to bring it about. Where some leave off, others take over.

How much there is still to learn is shown by the research of Schanderl concerning the nitrogen process in plants. It has been known for a long time that certain bacteria create small nodules on the roots of leguminous plants. It was believed that these bacteria had the rare ability of assimilating nitrogen from the air and giving it to the plant, thus making it independent of the nitrogen in the soil. Schanderl was able to show that nitrogen bacteria do not occur only on roots but in the whole plant, especially in its green parts, and not only in leguminosae but in a great number of plant species. In the green parts of plants bacteria are of a different shape and nature. They are nodular or thread-like integral parts of the protoplasm, but when isolated, they become independent organisms. They had been described earlier but their significance had not been understood.

Since this has become known we should not draw a dividing line between the living plant and the life of the soil. The plant emerges as the respiratory organ of the Earth and where the intake of nitrogen is concerned we see bacteria playing their part.

The bacteria draw together the processes of the whole plant into a single cell. The spore is a seed reduced to a minimum. The ability to produce coloured patches too must be considered in relation to the whole plant. Colours – red, yellow, white, etcetera – appear when bacteria are grown in pure cultures on gelatine nutrient or agar-agar. We see, therefore, that colouring appears even at this very lowest stage.

24 Mosses and Liverworts

Everybody knows this fascinating miniature world and yet it is veiled in mystery. Here we have a stage of the plant kingdom very limited in size and restricted to its own particular level. All mosses are dwarfs and their range of size is relatively small. The largest is to be found in New Zealand and reaches a height of 50 cm (20 in.). A giant among mosses! By contrast, what an enormous variation in size we have in the higher plants.

There are two forms: the mosses and the liverworts. We shall begin with the former. All of them make cushions. A single moss plant would

Plate 81: 'Flowers' of Common Bank Hair Moss (*Polytrichum commune*). This moss is dioecious. Only the male stalklets spread out their topmost leaves forming a little cup. The antheridia are placed at the bottom of the cup into which they empty their spermatozoids.

be an absurdity, for only in a mass, a carpet, are they a viable whole. The characteristic habitat of the mosses is the damp forest floor. In dry places, on sand and rocks, we find usually only very minute varieties. In the tropical mountain forests, however, mosses proliferate in unimaginable exuberance.

Only in the tundra do mosses have a certain influence on the appearance of the vegetation. One speaks of a moss tundra. Under the influence of the inimical arctic conditions even higher plants take on the appearance of mosses. They spread out at ground level and contract their shoots to form cushions. But mosses do not have such branch systems uniting in one root. For the mosses as for fungi, the Earth itself is the stem.

What would the Earth look like if she were only capable of moss growth? She would have only a very scant cover of vegetation – a kind of fur. There would be no trees. The vegetation would be like the tundra where the cosmic forces bring to life only the uppermost layer

Plate 82: *Polytrichum commune* with spore capsules. The 'Pixie Hoods' are the enlarged archegonia, torn off and carried up with the capsule as it stretches.

of the Earth and where the life of the Earth itself is reduced to a minimum. The mosses spread a tundra carpet all over the Earth and build a second plant world under and among the normal vegetation. Even in the tropics, mosses change their characteristics but little.

Mosses are closely related to the mineral earth. They are a kind of transition stage between dead mineral and live plant, as can be seen by those species which continue to grow above whilst becoming peat below. Sphagnum Moss (peat moss) leaves behind enormous layers of organic matter in the course of time. The carbon of the plant has become mummified, as it were, in the peat.

The mosses consist of little stems and leaves which make them look like minute trees or twigs of trees. Nature has taken great trouble over the making of them. Nothing is left vague or unfinished. It is wonderful to see. This again is a property which they have in common with arctic vegetation. (They would seem to belong to cold conditions.)

The moss plantlets stand on the soil without roots. Even root fibres (rhizoids) are often only there in the beginning; later on the plantlets support each other. Neither do they develop upwards. We speak of a 'flower', but what appears at the end of minute stalks are antheridia and archegonia. There is no true flower. Mosses are only middle sections of plants – green shoots. Like ferns, they propagate by means of spores. With the help of the microscope it was found that mosses also have two different phases of development. Strange to say, the little green plantlet corresponds to the prothallium in ferns, whilst the pixidium, the little capsule containing the spores, corresponds to the fern itself. These correspondences result from the number of chromosomes in the cell nucleus. The nucleii of the prothallium and the vegetative stage of the moss plantlets have only half the number of chromosomes the ferns or the moss spore capsules have. Through the fertilization of the ovum on the prothallium, or on the moss stalklets, the chromosome number is doubled. Only when the spores are produced is the number again halved. (So-called reduction division.) In both cases there are two quite different phases which succeed each other in a regular rhythm (alternation of generations). The dissimilarity arises out of a wrong comparison. The moss plantlet should be compared with the prothallium and not with the fern or equiesetum. The moss plantlet has a stem and leaves like a higher plant, yet it is only a

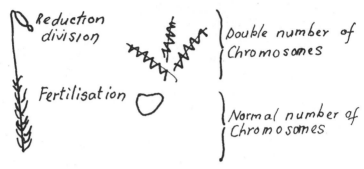

Fig. 34.

prothallium, differing in shape from the fern prothallium. It carries antheridia and archegonia.

Out of the fertilized ovum of the archegonia the spore capsule (*sporogonia*) grows on a thin stalk out of the crown of the moss plantlet, like the young fern plant out of its prothallium. It sits there like a parasite on its host. It is possible to pull out and put back the stalk bearing the capsule and even to transplant it on to another similar moss plantlet without disturbing the development of the spores. This shows that it is indeed an independent plant entity, even though it grows on a plant.

Out of the spore comes – not the prothallium again, for the moss plantlet is the prothallium – nor yet the leafy moss plantlet itself, but a fine thread-like web of green colour, the protonema, reminiscent of the thread-like algae. One might regard it as a brief recapitulation of the algal stage. The protonema has no organs of propagation and multiplies vegetatively by producing buds which grow into moss plantlets. The protonema of the luminous moss (*Hookeria lucens*) has achieved a certain fame as it is luminous. Its lens-shaped cells have the peculiar capacity of reflecting daylight and thus give it a gold-green sheen which appears almost magical in the twilight of dark places, caves and holes where it grows. The actual moss plantlet is not different from other mosses.

Looking at mosses again takes us far back into past periods of the Earth's evolution. It would be difficult to place them in the ascending scale of algae, ferns and flowering plants. Nature's attempt

Plates 83 and 84: Two pages from a collection of landscapes done in mosses, made by the moss specialist Adalbert Geheb, and presented to Ernst Haeckel. They were intended to illustrate the point that the German mosses display all the plant forms of the world in miniature. The upper picture represents an oasis, the lower one a Riviera landscape. (By permission of the Ernst Haeckel Archives in Jena).

simply to place the prothallus (the algal stage) on dry land and develop it directly into a flowering plant has failed. Only miniatures resulted – dwarf plants – incapable of further development. So the mosses too show us that an already highly developed stage (the alga corresponding to the prothallium) cannot rise to the next stage without fundamental transformation.

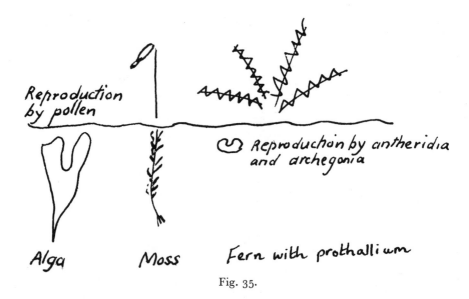

Fig. 35.

If the moss plantlet and fern prothallium correspond, how can the sporogonia (spore capsule) be interpreted? We must try to find an explanation in the higher plant, the flowering plant. The climax of development in the moss world is reached when something which is a continuation of seed capsule and anther is placed on top of the slender stalk. The spores of mosses are both seed and pollen grain combined.

In the mosses we find a most remarkable fusion of these two components which must bring to a halt all further phylogenetic development. In the moss plantlet we have an over-developed prothallium modelled on the higher plant, which however is only a product of the merging of seed capsule and anther.

At what moment in the Earth's development this caricature of synthesis took place, palaeontology is not yet able to say.

The liverworts show a much greater resemblance to algae than do

Plate 85: Sphagnum Moss.
(Enlarged).

mosses. Liverworts are not composed of stem and leaves, but a flattened thallus which is attached to the ground. Their dichotomic branching (dividing into two) makes them more like algae. Some used to be used as a remedy for liver complaints – hence the name. They are entirely leaf. It is interesting how the inherent forming forces in the liverworts appear when the thallus proceeds to form its antheridia and archegonia. One of the liverworts (*Marchantia polymorpha*) sets about its propagation in the following way: A little umbrella is produced which is simply a branch of the thallus. The stalk is a lengthened and narrow bit of thallus and the umbrella is star shaped. As this liverwort is bi-sexual, it has male and female umbrellas, the former bearing antheridia, the latter archegonia. The male umbrella has a slight resemblance to a mushroom. It has an inverted rim, and the resulting little bowl gets filled with rainwater or dew. The antheridia are placed at the bottom of the little bowl and the sperm cells are

Plate 86: Common Liverwort with little cups in which the buds develop and separate, thus furthering vegetative reproduction.

discharged into the water. Now the male sperms in their minute aquaria must find their way to the archegonia on the female umbrella. Here we find something reminiscent of the process in the flower of higher plants. The splashing rain sprays the cells on to the neighbouring umbrellas which, at a very primitive stage, are minute flowers. Liverwort flowers are not pollinated. They, like a prothallus, are watered. When fertilization has taken place, the spore capsules hang on the underside of the female umbrella. They are very small. This liverwort, by its method of reproduction in a *single* drop of water, makes the first attempt at individualizing reproduction by separating it from the totality of the Earth.

If we again compare the mosses with the liverworts we see that the mosses develop the stalk and seed-capsule tendency. The umbrella of the liverwort resembles a starry flower. The leafy type of liverwort, *Marchanta polymorpha*, can copy only the leafy part of the flower (sepals and petals).

186

Fig. 36: Left male, right female 'umbrellas' of Common Liverwort.

We must remember that the little umbrella of the liverwort is only a branch of the thallus, and therefore corresponds with the prothallus, while the spore capsule of the moss corresponds with the fern or the flowering plant. The liverwort, too, is an unsuccessful attempt at transferring the algal stage to dry land and open air. As the mosses appear, the stem tendency takes over – the upright posture. In the liverworts the upright tendency is only in the fruiting bodies. In the mosses it is their stalks which show the transitional stages from creeping to uprightness. According to Rudolf Steiner, the child at the age when it learns to stand up could be compared with the mosses.* Flower tendencies in anticipation can be detected, though they are of a very provisional character.

* See footnotes pp. 130, 149, and 167.

25 Lichens

Lichens settle in places unfit for any other plant growth. Bare stones and the bark of trees are their habitat. In the polar regions and high mountains lichens extend further than mosses ever could. Some form greyish-green or yellow crusts, others look more scaly, others again hang from trees in tangled confusion or single threads. Those resembling anthers or cups are specially attractive. With the exception of 'Iceland Moss' and 'Reindeer Moss' they rarely form a carpet. Some are lobate and faintly resemble liverworts, especially those growing in damp places. It is hardly possible to find a law which embraces all the great variety of lichens, nor does comparison with higher plants get us very far. The tolerance of lichens is almost unbelievable. They can live without water for many months and come to no harm even when

Plate 87: Lichen.

Plate 88: *Cladonia squa-mosa.*

drying out completely. If they then fall to pieces each piece can be the beginning of a new lichen. In fact their vegetative propagation relies on small fragments breaking off and finding a new place to settle in. The lichen does not break up into single cells but disintegrates into fragments. This sort of scattering is foreign even to fungi which have a special organ for casting off their conidiae. The connection between inner substance and outer form in these plants must differ very greatly from what it is in a green plant. One cannot imagine a green plant employing a mere formless crumbling-away as a means of vegetative propagation. In lichens the formative principle seems to work from without, not from within. Yet somehow it must be possible to find a way to understand this very unplant-like plant. How did Nature set about creating, maintaining and propagating such a primitive

Plate 89. Group of lichens.

Plate 90: *Peltigera canina*, (Dog Lichen), on sandstone.

organism which seems to have nothing in common with any normal plant growth?

The microscope, which has so extended the sphere of our observation, permits an insight into the 'technique' of Nature. It has revealed the astonishing fact that lichens are the result of two different plants forming one unit. A fungus and an alga together make a lichen – an amazing fact. This life-partnership has been called symbiosis, and rightly so; but one must bear in mind that the result is only *one* plant. The lichen cannot be explained by the study of either fungi or algae alone, for both are only the elements Nature uses for the creation of a third.

The fungus envelopes the alga. If the alga is thread-like, the resulting lichen will be thread-like too. In most instances, however, the lichen contains many single-celled pin-points of alga. In these the shape is not determined by the alga but is due to a higher over-ruling formative principle. 'The body is evolved by an ideal model, a form-giving plan, a mystical, independent "type".' (Novalis, *Fragments*.)

Plate 91: Reindeer Moss (*Cladonia rangiferina*).

191

Wherever lichens grow, neither fungi nor algae alone could exist, although it has been possible to breed both components separately under artificial conditions. Needless to say, the result was not a lichen but a fungus and an alga.

The alga provides the layers of tissue which assimilate carbon dioxide which the chlorophyll-deficient fungus is unable to do. In the leaf-like lichens we find the alga cells where the chlorophyll-bearing layers are to be found in a normal green plant.

In this strange world sometimes the leaf-stem tendency predominates, sometimes the fungus. Dog Lichen (*Peltigera*) is purely foliate, Reindeer Moss is stemmy and branched, Beard Moss (*Usuea barbata*) has stalks and threads, while the shape of the fructicose lichens stems undoubtedly from the world of fungi. How are we to explain these very peculiar plants? They are a world apart from normal plant life. Are they contracted, dwarfed and hardened root-like remnants of a far-distant past? Are they images of a moon landscape with its craters, ridges and broken crusts? Besides living on the tundra and high mountains, they are found on the bark of trees and on poor sand, clay and gravel soils – places that reproduce in miniature the conditions of the higher zones of the Earth where, with diminished earth-vitality, plant life is but tenuous. An organic synthesis of terrestrial and cosmic forces can be achieved as in higher plants, but it is only an external, superficial mixture of fungus and alga, the former representing the earthly, the latter the cosmic component.

Besides vegetative reproduction, lichens also increase by means of spores. When the lichen thallus grows a fruiting body to produce spores, the fungus element appears. The alga has no part in this. These wonderful, often brilliant red or yellow fruiting bodies really look like minute fungi. There is however no question of metamorphosis.

Fig. 37: Lichen with fungus-like fruiting bodies.

V Evolutionary Considerations

26

In the distant past of Earth evolution plant life was very different from what we see around us today. Different plants grew under different conditions. Although fossils, the documents of palaeontology, do not suffice to give us a complete picture, they do give valuable glimpses and allow us to draw certain conclusions. The following considerations will be confined to material brought to light by palaeontology and we will endeavour from this to understand how the plant world has developed.

From the study of fossils one finds that the plants preceding our present ones differed from them materially. The plant kingdom only gradually became what it is today. Out of a primitive state, by adding ever new elements, it became more and more perfect. Evolution makes the gradual development intelligible and shows how everything links up.

Careful study makes it possible to reconstruct the plants of, for instance, the coal forests – *Sigillariae* and *Lepidodendra* of gigantic size, various types of tree ferns, strange equisitae, etcetera. The present representatives of this former vegetation give us but a faint idea of them. *Lycopodiae*, for instance, were tall trees and have now degenerated to the state of insignificant little plants.

What makes the plant life of this far distant period of evolution so different is the fact that what we most appreciate in our present plants, the flower, is missing. The coal forests are flowerless. Nearly all their plants are at the fern stage – a soulless and monotonous vegetation. It stands to reason that the plant world went through an evolution which brought it step by step nearer perfection, until it reached its present state. Forever new forms were thrown up by nature whilst others disappeared. The process of plant formation through the ages is like the metamorphosis of *one* plant, although the different stages of development are characterized by different types of plants.

To get a first glimpse of this development let us remember what has

Plate 92: Impression of a fern frond on shale.

been said about the nature of ferns in earlier chapters: they develop mainly the vegetative shoot, which in ferns is the leaf, in horsetails the stem. The development is arrested there, although future stages are already anticipated. The spore formation is not wholly integrated into the green parts, but appears to be superficially attached to the underside of the frond. What a fundamental transformation is the higher plant with its flower-process culminating in seed formation! Everything must be transformed; the green shoot must cease to dominate and the impulse from below upwards has to be extinguished. Only thus can a new system of reproduction be rekindled in the plant from above downwards. The upper world, the world of light, with its messengers, the bees and butterflies, becomes intimately united with the plant. The insects take over pollination.

The higher plant emerged gradually as it detached itself from a more animal-like stage. Palaeontology speaks clearly of this development. Animal life appeared on Earth before plants did. Long before plants left any traces of their existence hundreds of animal forms had appeared. Then followed a vegetation of some sort of algae which even contributed to the formation of rock.

The first plant to grow on dry land is similar to the carboniferous flora and appears only a little earlier. The classical fern vegetation of the carboniferous system appears quite suddenly on the Earth and without any preliminaries. This strange fact has not been appreciated enough. Certain events in the spiritual part of the Earth must have resulted in the hardening and casting-off of the animal part of it; this was followed by the physical appearance of vegetation, in, as it were, a second wave of the casting out process. If anyone wishes to protest that the plant must have been there first in order to supply the animal with food, he forgets to keep an open mind for the possibilities of evolution. Present day conditions should not necessarily be projected into the past. What holds good today need not always have done so.

Looking over the whole development of the plant kingdom as palaeontology has laid it before us, we find remarkable breaks at certain points in the flow of evolution. Development along a certain line suddenly ceases. Nature divides the vast movements of her symphony by pauses. The fundamental significance of these pauses had not yet been sufficiently appreciated. It was thought that the gaps left by palaeontological findings were a result of their incompleteness and would be gradually filled. This hope has not been fulfilled, and more and more voices are raised in support of a theory which sees an importance in these rhythmical interruptions. In what follows I shall try to explain this.

All phylogenetic development is discontinuous. Leaps are made and gaps divide the different stages. The facts demonstrate this clearly. The materialistic principle of the continuity of substance and force applied to the history of evolution inevitably leads to contradictions. The crest of the first wave of the development of plants growing in soil is the carboniferous flora; however, with the end of the Palaeozoic Era, which for the plant kingdom lies between the Lower New Red Sandstone and the Permian Limestone period (see the geological table at the end of this chapter), this highly developed flora with its many

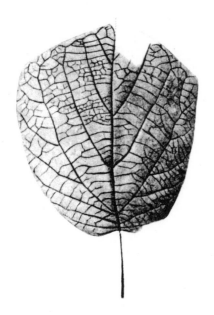

Plate 93: Credneria leaf from the chalk formation. Leaf with network of veins. (From *Grundlinien der Pflanzen-Morphologie im Lichte der Palaeontologie* by H. Potonie).

very distinct species has vanished almost completely. One could hardly have a more impressive fact than this. After the Triassic Red Sandstone period, which is characterized by its scanty plant growth, a new beginning is made: the flora of the Mesophytic Era. *The former vegetation, however, did not develop further.*

During the Mesophytic Era we again find highly developed plants of a special character, particularly in the Juarassic and Chalk formations. But this climax is also an end. In the Upper Chalk the rich variety of forms has disappeared. Suddenly, flowering plants spring up without warning, simultaneously in many different parts of the world. We need not violate palaeontological facts in order to find, for each of these great periods, one characteristic plant organ. In the carboniferous period it is the leafy vegetative shoot, corresponding to the fern. In the middle period (Mesophytic), with the predominance of conifers, Ginkgos and Cycads, the leaf-stalk type of plant has risen to the stage of seed bearing. In the next period, Upper Chalk and Tertiary, the real flowering plant finally appears.

With the new impulse, our present period (Neophytic) begins, and it gives the plant world again an entirely new appearance. Plants with conspicuous flowers like Magnolia, which are characteristic of the

196

Plate 94: Enlarged portion of a leaf of Lemon Balm (underside). It shows the netted mesh of veins of the flowering plant.

Tertiary flora, appear suddenly. Their rightful place is in between the conifers and the true flower-bearing plants, for the interior of the flower still bears reminiscences of the conifer stage. Only the large thick petals are a new addition. The magnolias tried to transform a conifer into a flowering plant, but overshot the mark.

A flowering plant is not merely a lower plant plus calix, corolla, etcetera. The higher flowering plants (*Angiosperms*) are an entirely new creation. New systems of branching appear, the veining of the leaves is different, tissues have a different structure and inner qualities change, heralding the new era.

At the beginning of the Neophytic Era we have the *Credneria-Leaf* as a symbol for the new principle which appears with the latest development. We do not know much about the plant to which it belonged. Some think it was something like a plane tree. The leaf is a true angiosperm leaf, but, already in the Upper Cretaceous period it has disappeared. Different shapes take its place and the vegetation becomes more and more like our present-day flora. The Tertiary period sees the end of the development of new types, though geographical changes go on, greatly altering the appearance of the face of the Earth.

Fossils reveal in the consecutive strata the sequence of development of the plant and animal kingdoms. The more primitive forms of lower organisms are found in the oldest strata, whilst in the younger strata an increasing perfection of types appears.

These palaeontological facts, bared of all theories, do no more than allow us to state that the lower organisms were developed first and were followed by the higher ones. To conclude – as the evolution theory does – that the lower forms of life *developed into the higher ones* means to get lost in theories and thereby violate the fundamentals of science. We have *proof* of the fact that the higher organisms *follow* after the lower ones, not that they descend from them.

Before continuing, it is necessary to dwell on certain characteristic elements of the carboniferous system. The enormous ferns of that period and the forests of gigantic horsetails as well as the highly developed clubmosses, are represented today by insignificant images of their former splendid selves. It is difficult to imagine that the forbears of the small creeping clubmosses were trees. (*Sigilariae* and *Lepidodenra*.) How strange these trees must have looked! Prints of their bark and fossilized stems are found in large numbers in coal mines. Their surfaces are very well preserved. Below the leafy crown emerged bunches of small cones from stems and branches, in Lepidodendra, sometimes at the ends of the branches. The cones can be likened to those of conifers, and also slightly resemble ears of corn. Hidden by the scales, however, were no seeds, but spores instead. These presumably formed prothallia as the spores of our present day clubmosses do, although the fossils found do not bear witness to the fact. *In these plants there is the gesture of a seed-bearing plant but the inner process is completely fern-like. These ferns conform to a law which becomes intelligible only in a later period of evolution.*

The next step is that some clubmoss trees differentiated between their spores and produced large ones (*megaspores*) and smaller ones (*microspores*). The former would correspond with the female, the latter with the male sex. In the prothallus of our ferns the ovum (*archegonium*) and the sperm (*antherozoid*) are still united. Horsetails have already two different prothallia. In the fossil clubmoss trees, the contrast of male and female spores is already clearly defined. Micro- and megaspores can be found in one common cone. The larger megaspores sit at the base of it. In some Lepidodendra the separation of sexes goes further.

Male and female cones are developed, as in our conifers. It goes without saying that this is a first step towards seed formation. We need only imagine that the megaspore settles in its place and sprouts into a prothallus while still attached to the scale of the cone, and we have a kind of seed. It is a wonderful process, and the gradual transformation from fern to flowering plant seems a convincing fact. Nevertheless, when the peak of development is reached, there is a sudden break. There is no further continuation.

At this stage something is foreshadowed which is only achieved in the future. The most highly developed clubmoss trees anticipate the seed-bearing plant which appears only in Mezophytic times, suddenly, all over the Earth, when the fern period has ended. Among the giant equisetae of the Carboniferous period were some with both small and large spores and, to the great surprise of scientists, it was found that some ferns had actually developed true seeds. However, long before this could be proved, the enormous number of seeds found had aroused great interest.

The great riddle remains: why is it that of all these highly developed Sigilaria, Lepidodendra, giant Equisetae and seed-bearing ferns, not a trace remains in the following period – the Permian – even though the beginning of this next stage of development is detectable in them? Not a single species survived. What deep secret lies hidden here?

Since the time of Haeckel we have become accustomed to the idea that an organism as it develops recapitulates briefly the history of its species. 'Ontogenesis is a short recapitulation of phylogenesis'. This fact, known as the biogenetic basic law, clarifies a great deal that the development of an individual would not reveal. In the development of the plant and animal kingdoms we continually come up against facts which are comprehensible only if seen as repetitions of phylogenesis. The biogenetic law points to the past. What Haeckel found valid for organisms clearly also holds good for geological periods.

Stimulated by Rudolf Steiner's ideas, let us reverse and enlarge the biogenetic law so that it not only embraces the past but also the future. Rudolf Steiner made the surprising discovery that in addition to a repetition of past developments, future developments are hinted at and anticipated. This means that a stage of evolution, having passed its climax, overreaches itself and becomes 'overripe'. The impulse for the development of the next stage must be already implanted so that it can

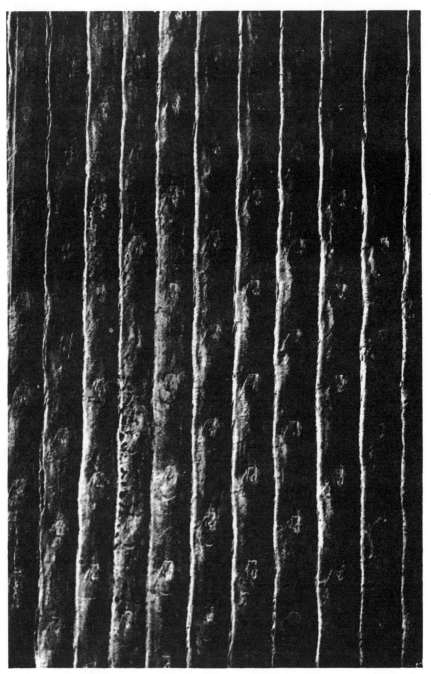

Plate 95: Carbonized bark of a *Sigilaria* from the Lugau-Oelnitz coalfield. Note the oblique lines of the leaf scars.

break through in the existing untransformed material, seeking to find expression. Organic evolution does not entail only progressive transformation of *forms*, but the very substance must be developed from stage to stage in order to create the conditions suitable for a certain level of organization. Transformation processes, like the insect metamorphosis from egg to fully developed insect, would remain unexplained but for this fact. There must be an appropriate base on which a new impulse can work. Yet it can happen that a higher form is impressed on a material which belongs to a lower stage and does not correspond with it.

Do not these ideas of Rudolf Steiner explain many of the riddles of palaeontology, some of which we have been considering? They can be called *prophetic forms*. Leo S. Berg uses this term in his book *Nomogenesis* with reference to certain developments in the animal kingdom, calling them 'phylogenetic acceleration or procession of phylogeny by ontogeny'.

The propagation of some of the clubmoss trees and Calamites by spores, and the seed formation of some of the ferns are anticipations of a later stage of evolution. *They are prophetic plant forms.*

Their rich variety is shown by B. A. C. Seward in his book *Fossil Plants* (Vol. 3, pp. 303, 304).

> Among the numerous types of Palaeozoic seeds are several which invite comparison with the fruits or carpels, apart from the seeds, of Angiosperms. Impressions of Samaropsis seeds bear a close resemblance to the laterally expanded fruits of the common Crucifer, *Thlaspi arvense*; the ribbed testa of Hexagonocarpus and other genera recalls the fruit-wall of Alstroemeria; the recently described lower carboniferous seed Thysanotesta sagittula, (Nath.) simulates a carpel of Erodium. These and similar instances of a close parallelism in external features between organs that are not homologous, though in themselves of no morphological significance, are at least interesting as illustrating the plasticity displayed by reproductive structures, which in the Palaeozoic period marked a morphological achievement comparable in its importance with the still greater achievement represented by highly specialized fruits of the modern flowering plants. The range in form and surface-features of Angiospermous fruits was foreshadowed by Palaeozoic seeds.

Let us repeat again: the seeds of those plants of the Carboniferous period take on only the outer appearance of fruits of Angiosperms. Their inner nature is quite different. The future state is imprinted on

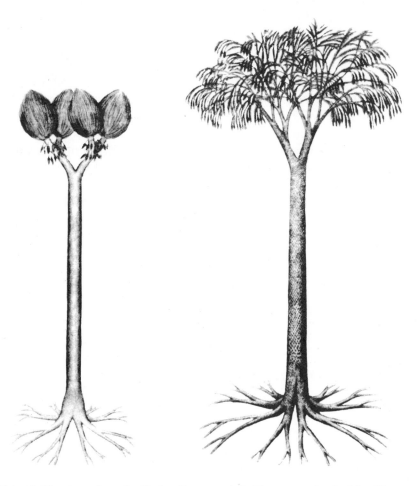

Plate 96: Tree types from the Carboniferous period. (Reconstruction by Max Hirmer, *Handbuch der Palaeobotanik*). Both trees have dicotomic branching, also in the roots. The *Sigilaria* (left) makes a strange impression. Its 'crown' consists of many bunches of thong-like leaves. On the trunk can be seen the scars of the fallen leaves arranged in spirals. Below the leafy crown hang the spore cones emerging directly from the stem. The height of the tree is about 10 metres. The spore cones of the *Lepidodendron* hang on the end of the leaf-clad branches. Height up to 15 metres.

unsuitable material and this is one reason why it cannot develop further.

The flora which follows this early period develops from new beginnings. The old does not proceed further. To ask how it is possible for so highly developed a flora as that of the coal measures to die out is to look at the problem in the wrong light. It is just *because* this flora was so highly developed, in fact overreaching itself, that it had to perish. It had lost its capacity for transformation. The appearance of seeds and the anticipation of later forms of fruit is no dawn, but rather an evening twilight. The night must fall before new young life capable of development can appear on the scene.

Again towards the end of the middle period (Mesophytic) prophetic forms appear. Instead of the spore and seed formation of the coal measures, it is now the flower (*corolla*) of the higher plants that is foreshadowed. This anticipation of the flowering plant is one of the most interesting developments that palaeontology has brought to light.

The indication of the division between the Mesophytic and Neophytic periods occurs in the middle of the chalk formations. (See geological table at the end of the chapter.) In the Upper Cretaceous layer flowering plants (Angiosperms) appear suddenly, almost unheralded. The two main groups, monocotyledons and dicotyledons arose probably at the same time. They appear simulatneously in different geographical localities. 'The palaeontological findings and studies of plant distribution let us assume that various families of Angiosperms started to develop simultaneously in different regions of the earth.' (Menzel.)

The new forms of life are in their own way fully developed when they appear. There are no transitional forms leading from the plants of the Mesophytic to those of the Upper Cretaceous period. In the Jurassic and Lower Chalk formations the Cycads, a kind of palm-fern, play an important part. They are Gymnosperms like our present day conifers. Their pinnate leaves, arranged in rosettes, give the plant a character all its own. Cycadophytes are seed-bearing plants. In most of the present-day ones the seeds are placed on the scales of the cones, often of considerable size, which grow directly out of the stem at the bases of the leaves. Whoever sees the descendants of these Mesophytic Cycadophytes, Cycads or Dioons, in fruit in a greenhouse, recognizes immediately that these are plant forms whose heyday was in far distant times.

Brilliant research, particularly by Professor Wieland, has made it possible to reconstruct the early Cycadophytes in detail, especially the well known but now extinct fossil group of the Bennettitales. The

Plate 97: Giant Horsetail (*Calamite from the group of Eucalamites*). (Reconstruction by Max Hirmer, *Handbuch der Palaeobotanik.*) Height up to 10 metres. The Giant Horsetail of the coal measures, unlike our present-day *Equisitae*, had a cambium layer and grew in thickness like our trees.

picture on page 208 shows what these flowers looked like. The 'pollen-leaves' which form the flower crown resemble fern fronds. The fern leaf is simply transplanted into the flower without any metamorphosis – a 'flower' of fern fronds, an 'improvised' flower. The pollen-leaves (really microsporophylli) unroll from the centre of the flower as it opens. Again this is a prophetic form, an abortive attempt to anticipate the flowering plant. The inevitable sequel was that the whole family of the Mesophytic Bennettiteae died out completely; they could not possibly live on in the era of the true flowering plant.

Science has not given up hope of one day finding transitional forms between the Bennettiteae and the true flowering plants. We think we have shown why this cannot happen. Bennettiteae are *forerunners* of true flowering plants, not *ancestors*. They die out, and evolution makes no further use of them. The new development starts from a point that we cannot find from palaeontological material.

W. Gothan writes in his textbook on palaeobotany:

> It must, however, be said that the removal of disagreements over the time of the appearance of Angiosperms has not in any way minimized the riddle of their sudden emergence in great numbers. Even though we must see the Bennettiteae as a kind of 'premonition' of the Agiosperms which they directly preceded in the Lower Cretaceous period, we are still far from knowing any transitional forms. Bennettiteae and dicotyledons for a time existed side by side, as the finds in the South of England show; the expectation, however, of finding in these strata the transitional forms, which must have been there, has so far not been realized.

Transitional forms are not to be expected. There is no continuity between the two groups of plants, even though they co-existed for a time. On page 437 in the same book Gothan writes: 'In the case of Cycaoidea maxima, the largest known Bennettitea cone containing 600 seeds, we notice that the appearance of enormous size precedes extinction.' Another case of overreaching development.

Let us sum up. Regarding plant evolution as palaeontology shows it, it cannot be a foregone conclusion that the higher is descended from the lower. Evolution is characterized by large periods separated from each other. Palaeobotany calls them Palaeophytic, Mesophytic and Cainophytic. Only *within* these periods can one speak of direct descent. The impulse for the next period is not a hereditary one. It is something entirely new. At the end of each period prophetic forms appear. They are old, and no longer pliable material 'mirrors' the new impulse. Prophetic forms are not hereditary sources of new forms but fore-shadowings. They cannot be explained from the past, but point to the future which in time will become reality. This is a revolutionary idea in relation to current theories.

In his book, *A Theory of Knowledge based on Goethe's World Conception** Rudolf Steiner says:

> Friedrich Theodor Vischer once expressed the opinion in regard to the Darwinian theory that it would render necessary a revision of our concept of time. Here we have arrived at the point which makes manifest to us in what sense such a revision would have to occur. It would have to show that the deducing of a later from an earlier is no explanation; that the

*Anthroposophic Press, New York, 1968.

first in time is not the first in principle. Every deviation must originate in this first principle, and at most it would be necessary to show what factors were effective in bringing it about that one sort of entity evolved in time before another.

If one applies the theory of descent without qualification to all evolution, then the existence of prophetic forms must be ignored. But this gives a distorted picture of phylogenetic evolution. One cannot make sense of the gaps in the ascending line, and one looks in vain for transitional forms where none can possibly be found.

The true impulses of progressive evolution cannot be discovered by studying the past, for they are of a spiritual nature. Unfolding creation leaves only its foot-prints in the sensible world. Since this point of view first appeared in print, new observations and expressions such as 'anticipations', 'foreshadowing imitations', etcetera, have crept into modern palaeontology, for the facts speak unmistakably for themselves.

To close this book the author would like to dwell on a point which he believes will influence the philosophy of evolution. The 'first cause' responsible for the progressive advance of natural evolution can be conceived of only as a spiritually active principle. Here one must not confuse new creations appearing at particular turning points with original primary creation. When an undisturbed, continuous development covering a long period suddenly makes a leap in a direction which cannot be explained from the past (such as when the Phanerogams suddenly appeared on the Earth), it is nevertheless comprehensible that the new impulse takes hold of material that is still malleable enough to respond, and in a very short time raises it to a higher stage. This gives the impression that new forms emerge for no apparent reason. These can well be called 'new creations', for the new type cannot be explained from the old material in which it has clothed itself. We find such leaps everywhere in nature since all development is *discontinuous*, even in the individual plant which, as we pointed out, is a picture of phylogenetic evolution. Compared to the green shoot, is not

Plate 98: *Cicas revoluta*. Female plant in flowering stage. Our present day cycadas, unlike most fossils, are dioicious: male and female flowers are borne on different plants. Our illustration shows the flower in the heart of the palm-like plant shortly before unfolding. The seeds develop on divided carpel leaves. There is no true corolla. The male flowers are cone-like. (Photographed in Dalmatia).

Plate 99: Reconstruction of a *Bennettitales* flower (Glass model designed by Prof. Wieland). Though the *Bennettitales* are classed among the gymnosperms like our conifers. nevertheless they formed a sort of corolla which consisted of fern-like, branched anthers. (*Microsporophyllae*). The cone-like arrangement of fruit is in the centre of the flower and is not visible in the illustration. The flower emerged directly from the stems of the palm-like plants.

the flower a new creation? The leap should not be overlooked, for no one could predict the flower who had not already seen one. As the plant is preparing for the flower the latter is in no wise physically present in the green shoot. As in the microcosm so in the macrocosm.

A great deal would be gained if our scientific world conception would allow for the development of the physical from the spiritual. Facts seem to demand a recognition of the spiritual; we need add nothing, only let them speak for themselves and interpret themselves. A break-through to the spiritual in natural science could be achieved if conventional, materialistic habits of thought could be discarded – they cloud our perception. Pure observations could then speak for themselves.

Geological Table

Archaeic	Remains of organisms doubtful.	

Palaeo-phytic	Cambrian Silurian	Marine invertebrates and algae. Corals, Trilobites, Brachiopods, first fishes, seaweeds and first land-plants (Psilophytes).	Palaeozoic
	Devonian	Corals, panzer armoured fishes, and the predecessors of our conifers.	
	Carboniferous	Coal deposits, highest development of ferns (seed ferns) scale trees (Lycopods) giant horsetails, first true conifers, spiders, insects and amphibians. Trilobites die out.	
	Lower New Red Sandstone	First reptiles, treeferns and conifers (Araucarieas).	
Meso-phytic	Permian Limestone	Rock salt, marl slate, flora scant (Ginkgo).	Mesozoic
	Triassic	1. Red and white sandstone, formation of deserts with few plants (Horsetails, conifers). 2. Shell limestone. 3. Shales, crocodiles.	
	Jurassic	Highest development of Ammonites and Belemnites, vertebrate fishes, gigantic Saurians, reef building corals, birds, palmferns together with conifers.	
	Lower Cretaceous	Highly developed palmferns.	
Caeno-phytic	Upper Cretaceous	Belemnites and Ammonites die out. So do palm ferns. First Angiosperms (deciduous trees, Credneria).	Caenozoic
	Tertiary	Great upheavals through faults, volcanic eruptions, etcetera. Gigantic mammals, hoofed animals, monkeys and apes. Many Angiosperms (elm, birch, palm, etcetera). All present day plant families are represented. No further new forms are developed (Brown coal).	
	Quarternary (with ice age)	Man appears in Europe. From now on flora and fauna undergo only geographic changes.